Power System Analysis

Mehdi Rahmani-Andebili

Power System Analysis

Practice Problems, Methods, and Solutions

 Springer

Mehdi Rahmani-Andebili
State University of New York
Buffalo, NY, USA

ISBN 978-3-030-84769-2   ISBN 978-3-030-84767-8   (eBook)
https://doi.org/10.1007/978-3-030-84767-8

This Springer imprint is published by the registered company Springer Nature Switzerland AG
The registered company address is: Gewerbestrasse 11, 6330 Cham, Switzerland

Preface

Electric Power System Analysis is one of the fundamental courses of Electric Power Engineering major which is taught for junior students. The subjects include fundamental concepts in power system analysis, transmission line parameters, transmission line model and performance, modeling of power system components, and determination of network impedance and admittance matrices, load flow, and economic load dispatch.

Like the previously published textbooks, this textbook includes very detailed and multiple methods of problem solutions. It can be used as a practicing textbook by students and as a supplementary teaching source by instructors.

To help students study the textbook in the most efficient way, the exercises have been categorized in nine different levels. In this regard, for each problem of the textbook a difficulty level (easy, normal, or hard) and a calculation amount (small, normal, or large) have been assigned. Moreover, in each chapter, problems have been ordered from the easiest problem with the smallest calculations to the most difficult problems with the largest calculations. Therefore, students are suggested to start studying the textbook from the easiest problems and continue practicing until they reach the normal and then the hardest ones. On the other hand, this classification can help instructors choose their desirable problems to conduct a quiz or a test. Moreover, the classification of computation amount can help students manage their time during future exams and instructors give the appropriate problems based on the exam duration.

Since the problems have very detailed solutions and some of them include multiple methods of solution, the textbook can be useful for the underprepared students. In addition, the textbook is beneficial for knowledgeable students because it includes advanced exercises.

In the preparation of problem solutions, it has been tried to use typical methods of electrical circuit analysis to present the textbook as an instructor-recommended one. In other words, the heuristic methods of problem solution have never been used as the first method of problem solution. By considering this key point, the textbook will be in the direction of instructors' lectures, and the instructors will not see any untaught problem solutions in their students' answer sheets.

The Iranian University Entrance Exams for the Master's and PhD degrees of Electrical Engineering major is the main reference of the textbook; however, all the problem solutions have been provided by me. The Iranian University Entrance Exam is one of the most competitive university entrance exams in the world that allows only 10% of the applicants to get into prestigious and tuition-free Iranian universities.

Butte, MT, USA   Mehdi Rahmani-Andebili

Contents

About the Author

Mehdi Rahmani-Andebili is an Assistant Professor in the Electrical Engineering Department at Montana Technological University, MT, USA. Before that, he was also an Assistant Professor in the Engineering Technology Department at State University of New York, Buffalo State, NY, USA, during 2019–2021. He received his first M.Sc. and Ph.D. degrees in Electrical Engineering (Power System) from Tarbiat Modares University and Clemson University in 2011 and 2016, respectively, and his second M.Sc. degree in Physics and Astronomy from the University of Alabama in Huntsville in 2019. Moreover, he was a Postdoctoral Fellow at Sharif University of Technology during 2016–2017. As a professor, he has taught many courses such as Essentials of Electrical Engineering Technology, Electrical Circuits Analysis I, Electrical Circuits Analysis II, Electrical Circuits and Devices, Industrial Electronics, Renewable Distributed Generation and Storage, and Feedback Controls. Dr. Rahmani-Andebili has more than 100 single-author publications including textbooks, books, book chapters, journal papers, and conference papers. His research areas include Smart Grid, Power System Operation and Planning, Integration of Renewables and Energy Storages into Power System, Energy Scheduling and Demand-Side Management, Plug-in Electric Vehicles, Distributed Generation, and Advanced Optimization Techniques in Power System Studies.

Abstract

In this chapter, the problems concerned with the fundamental concepts of power system analysis are presented. The subjects include phasor representation of signals, voltage and current in power system, impedance and admittance, single-phase and three-phase power systems, complex power and its components, power generation and consumption concepts, per unit (p.u.) system, and power factor correction. In this chapter, the problems are categorized in different levels based on their difficulty levels (easy, normal, and hard) and calculation amounts (small, normal, and large). Additionally, the problems are ordered from the easiest problem with the smallest computations to the most difficult problems with the largest calculations.

1.1. What is the phasor representation of the voltage signal of $\sqrt{2}\cos(t)$?

Difficulty level ● Easy ○ Normal ○ Hard
Calculation amount ● Small ○ Normal ○ Large

1) $1\ V$
2) $(1\angle 90°)\ V$
3) $0\ V$
4) $(1\angle -90°)\ V$

1.2. Represent the current signal of $\sqrt{2}\sin(t)$ in phasor domain.

Difficulty level ● Easy ○ Normal ○ Hard
Calculation amount ● Small ○ Normal ○ Large

1) $1\ A$
2) $(1\angle 90°)\ V$
3) $0\ A$
4) $(1\angle -90°)\ V$

1.3. Define the signal of $\cos(2t + 30°)$ in phasor domain.

Difficulty level ● Easy ○ Normal ○ Hard
Calculation amount ● Small ○ Normal ○ Large

1) $1\angle 30°$
2) $2\angle -30°$
3) $\frac{1}{\sqrt{2}}\angle 0°$
4) $\frac{1}{\sqrt{2}}\angle 30°$

1.4. Represent the signal of $10\sin(t - 60°)$ in phasor form.

Difficulty level ● Easy ○ Normal ○ Hard
Calculation amount ● Small ○ Normal ○ Large

1) $10\angle{-150°}$
2) $10\angle{-60°}$
3) $5\sqrt{2}\angle{-150°}$
4) $10\angle{60°}$

1.5. In the single-phase power system of Fig. 1.1, the voltage and current are as follows:

$$v(t) = 110\cos\left(\omega t + 30°\right) V$$

$$i(t) = 0.5\cos\left(\omega t - 30°\right) A$$

Determine the impedance, resistance, and reactance of the system seen from the beginning of the line.

Difficulty level   ● Easy   ○ Normal   ○ Hard
Calculation amount   ● Small   ○ Normal   ○ Large

1) $(220\angle{-60°})\,\Omega, 110\,\Omega, 110\sqrt{3}\,\Omega$
2) $(220\angle{60°})\,\Omega, 110\,\Omega, 110\sqrt{3}\,\Omega$
3) $(220\angle{-30°})\,\Omega, 110\sqrt{3}\,\Omega, 110\,\Omega$
4) $(220\angle{30°})\,\Omega, 110\sqrt{3}\,\Omega, 110\,\Omega$

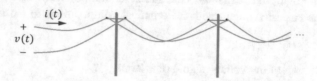

Fig. 1.1 The power system of problem 1.5

1.6. In the single-phase power system of Fig. 1.1, the voltage and current are given as follows:

$$v(t) = 100\sqrt{2}\cos\left(t\right) V$$

$$i(t) = \sqrt{2}\cos\left(t - 30°\right) A$$

Determine the admittance, conductance, and susceptance of the system seen from the beginning of the line.

Difficulty level   ● Easy   ○ Normal   ○ Hard
Calculation amount   ● Small   ○ Normal   ○ Large

1) $(0.01\angle{30°})\,\Omega, 0.005\,\Omega, 0.005\sqrt{3}\,\Omega$
2) $(0.01\angle{30°})\,\Omega, 0.005\sqrt{3}\,\Omega, 0.005\,\Omega$
3) $(0.01\angle{-30°})\,\Omega, 0.005\sqrt{3}\,\Omega, 0.005\,\Omega$
4) $(0.01\angle{-30°})\,\Omega, 0.005\,\Omega, 0.005\sqrt{3}\,\Omega$

1.7. The impedance of a generator, with the rated specifications of 20 kV and 200 MVA, is $\mathbf{Z} = j0.2\ p.\,u.$ Determine its reactance in percent if 21 kV and 100 MVA are chosen as the base voltage and power.

Difficulty level   ● Easy   ○ Normal   ○ Hard
Calculation amount   ● Small   ○ Normal   ○ Large

1) 11%
2) 10.5%
3) 11.7%
4) 9.07%

1.8. The reactance of a generator, with the nominal specifications of 14 kV and 500 MVA, is 1.1 $p.$ $u.$ Determine its impedance in percent if 20 kV and 100 MVA are chosen as the base voltage and power.

Difficulty level   ● Easy   ○ Normal   ○ Hard
Calculation amount   ● Small   ○ Normal   ○ Large

1) 30.8%
2) 10.78%
3) 60.8%
4) 57.8%

1.9. In the power bus of Fig. 1.2, determine the $i_3(t)$ if we know that $i_1(t) = 10 \cos(10t)$ A, $i_2(t) = 10 \sin(10t)$ A, and $i_4(t) = 10\sqrt{2} \cos(10t + 45°)$ A.

Difficulty level   ● Easy   ○ Normal   ○ Hard
Calculation amount   ○ Small   ● Normal   ○ Large

1) $10\sqrt{2}$ A
2) $(5\angle-45°)$ A
3) $(10\angle45°)$ A
4) 0 A

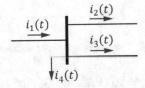

Fig. 1.2 The power system of problem 1.9

1.10. In the single-phase power bus of Fig. 1.3, $V_{rms} = 200$ V and the equivalent impedance of the loads are $Z_1 = (8 - j6)\ \Omega$ and $Z_2 = (3 + j4)\ \Omega$. Calculate the total active power consumed in the bus.

Difficulty level   ○ Easy   ● Normal   ○ Hard
Calculation amount   ○ Small   ● Normal   ○ Large

1) 8 kW
2) 15 kW
3) 7.5 kW
4) 9 kW

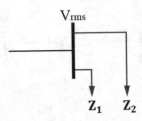

Fig. 1.3 The power system of problem 1.10

1.11. Calculate the instantaneous power of a single-phase power system that its voltage and current are $v(t) = 110\sqrt{2} \cos(120\pi t)$ V and $i(t) = 2\sqrt{2} \cos(120\pi t - 60°)$ A.

Difficulty level   ○ Easy   ● Normal   ○ Hard
Calculation amount   ○ Small   ● Normal   ○ Large

1) 110 W
2) $220 \cos(240\pi t - 60°)W$
3) $55 + 110 \cos(240\pi t - 60°)$ W
4) $110 + 220 \cos(240\pi t - 60°)$ W

1.12. In the single-phase power system of Fig. 1.4, calculate the active and reactive powers transferred from bus 1 to bus 2. Consider the following data:

$$\mathbf{V_1} = (10\underline{/30°})\,V, \mathbf{V_2} = (5\sqrt{3}\underline{/0°})\,V, \mathbf{Z} = j5\,\Omega$$

Difficulty level   ○ Easy   ● Normal   ○ Hard
Calculation amount   ○ Small   ● Normal   ○ Large

1) $10\,W, 10\sqrt{3}\,VAr$
2) $5\,W, 4\sqrt{3}\,VAr$
3) $5\,W, 4\,VAr$
4) $5\sqrt{3}\,W, 5\,VAr$

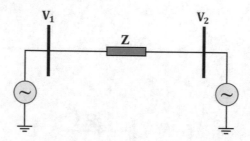

Fig. 1.4 The power system of problem 1.12

1.13. In the power system of Fig. 1.5, $\mathbf{E_1} = 200\,\underline{/-30°}\,V, \mathbf{E_2} = 200\underline{/0°}\,V, \mathbf{Z} = j5\,\Omega$. Which one of the following choices is true?

Difficulty level   ○ Easy   ● Normal   ○ Hard
Calculation amount   ○ Small   ● Normal   ○ Large

1) The first electric machine is generating reactive power, and the second electric machine is consuming reactive power. Moreover, the first and the second electric machines are working as a generator and a motor, respectively.
2) The first electric machine is consuming reactive power, and the second electric machine is generating reactive power. Moreover, the first and the second electric machines are working as a motor and a generator, respectively.
3) Both electric machines are generating equal reactive power which is consumed by the reactance of the line. Moreover, the first and the second electric machines are working as a generator and a motor, respectively.
4) Both electric machines are generating equal reactive power which is consumed by the reactance of the line. Moreover, the first and the second electric machines are working as a motor and a generator, respectively.

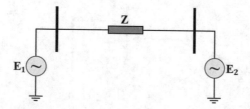

Fig. 1.5 The power system of problem 1.13

1.14. In the power bus of Fig. 1.6, the base voltage and power are 20 kV and 100 MVA, respectively. If a reactor is connected to this bus, determine its reactance in per unit (p.u.).

Difficulty level   ○ Easy   ● Normal   ○ Hard
Calculation amount   ○ Small   ● Normal   ○ Large

1) 0.25
2) 0.5
3) 0.75
4) 2

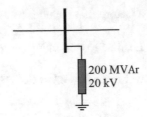

Fig. 1.6 The power system of problem 1.14

1.15. Figure 1.7 shows the single-line diagram of a power system with the following specifications. Calculate the resistance of the load in per unit (p.u.) if the nominal quantities of the generator are chosen as the base quantities:

$$G : 20 \, kV, 300 \, MVA$$

$$T_1 : 20/200 \, kV, 375 \, MVA$$

$$T_2 : 180/9 \, kV, 300 \, MW$$

$$Load : 9 \, kV, 180 \, MW$$

Difficulty level   ○ Easy   ● Normal   ○ Hard
Calculation amount   ○ Small   ● Normal   ○ Large
1) 1.25 p. u.
2) 1.35 p. u.
3) 1.45 p. u.
4) 1.55 p. u.

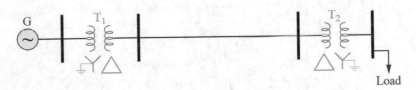

Fig. 1.7 The power system of problem 1.15

1.16. Figure 1.8 illustrates the single-line diagram of a power system with the given information. Calculate P and Q in per unit (p.u.). In this problem, assume that $\sin(15^\circ) \equiv 0.25$ and $\cos(15^\circ) \equiv 0.96$.

Difficulty level   ○ Easy   ● Normal   ○ Hard
Calculation amount   ○ Small   ● Normal   ○ Large
1) $P = 0.5 \, p. u., Q = 0.08 \, p. u.$
2) $P = 0.8 \, p. u., Q = 0.5 \, p. u.$
3) $P = 0.8 \, p. u., Q = -0.5 \, p. u.$
4) $P = 0.5 \, p. u., Q = -0.08 \, p. u.$

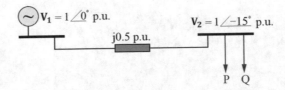

Fig. 1.8 The power system of problem 1.16

1.17. Calculate the complex power delivered to a factory that includes two loads with the following specifications:

$$\text{Inductive Load}: P_1 = 60\, kW, Q_1 = 660\, kVAr$$

$$\text{Capacitive Load}: P_2 = 240\, kW, PF = 0.8$$

Difficulty level   ○ Easy   ● Normal   ○ Hard
Calculation amount   ○ Small   ● Normal   ○ Large
1) $(180 + j840)\, kVA$
2) $(300 + j480)\, kVA$
3) $(300 + j840)\, kVA$
4) $(180 + j480)\, kVA$

1.18. Figure 1.9 shows the single-line diagram of a balanced three-phase power system, in which a synchronous generator has been connected to a no-load transmission line through a transformer.
Calculate the Thevenin reactance seen from the end of the transmission line. In this problem, the rated quantities of the generator are considered as the base values:

$$G: 20\, kV, 300\, MVA, X_G = 20\%$$

$$T_1: 20/230\, kV, 150\, MVA, X_T = 0.1\, p.u.$$

$$\text{Line}: 176.33\, km, X_{Line} = 1\, \Omega/km$$

Difficulty level   ○ Easy   ● Normal   ○ Hard
Calculation amount   ○ Small   ● Normal   ○ Large
1) $0.9\, p.\, u.$
2) $1.2\, p.\, u.$
3) $1.3\, p.\, u.$
4) $1.4\, p.\, u.$

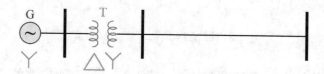

Fig. 1.9 The power system of problem 1.18

1.19. For the three-phase power system of Fig. 1.10, the following specifications have been given. Determine the voltage drop of the line in percent:

$$\text{Line}: \mathbf{Z} = (10 + j40)\, \Omega/phase$$

$$\text{Load}: V = 100\, kV, S = 50\, MVA, PF = 0.8\, Lagging$$

Difficulty level   ○ Easy   ● Normal   ○ Hard
Calculation amount   ○ Small   ● Normal   ○ Large
1) 8%
2) 16%
3) 19%
4) 24%

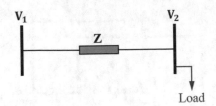

Fig. 1.10 The power system of problem 1.19

1.20. In the power system of Fig. 1.11, calculate the impedance of the load in per unit (p.u.) for the following specifications. In this problem, 20 kV (in the generator side) and 3 MVA are chosen as the base voltage and power:

$$G : 20\,kV, 3\,MVA, 3\%$$

$$T_1 : 20/230\,kV, 3\,MVA, 5\%$$

$$T_2 : 230/11\,kV, 3\,MVA, 5\%$$

$$Load : 11\,kV, 0.2\,MVA, 0.8\,Lagging$$

$$M : 11\,kV, 1\,MVA, 5\%$$

$$C : 0.5\,MVA$$

Difficulty level   ○ Easy   ● Normal   ○ Hard
Calculation amount   ○ Small   ● Normal   ○ Large

1) $(12 + j9)\,p.\,u.$
2) $(18 + j15)\,p.\,u.$
3) $(15 + j12)\,p.\,u.$
4) $(12.75 + j7.9)\,p.\,u.$

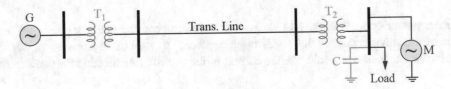

Fig. 1.11 The power system of problem 1.20

1.21. In the single-phase power bus of Fig. 1.12, the characteristics of the loads are as follows. Determine the total power factor of the bus:

$$Load\,1 : P_1 = 25\,kW, Q_1 = 25\,kVAr$$

$$Load\,2 : S_2 = 15\,kVA,\ \cos(\theta_2) = 0.8\,Leading$$

$$Load\,3 : P_3 = 11\,kW,\ \cos(\theta_3) = 1$$

Difficulty level   ○ Easy   ● Normal   ○ Hard
Calculation amount   ○ Small   ● Normal   ○ Large

1) 0.94 Lagging
2) 0.94 Leading
3) 0.6 Lagging
4) 0.6 Leading

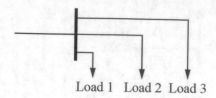

Fig. 1.12 The power system of problem 1.21

1.22. In the single-phase power bus of Fig. 1.13, determine the capacitance of the shunt capacitor that needs to be connected to the bus to adjust its power factor at one for the following data:

$$Load : S = 20\ kVA, \cos(\theta) = 0.8\ Lagging$$

$$V_{rms} = 200\ V, f = 50\ Hz, \pi \cong 3$$

Difficulty level   ○ Easy   ○ Normal   ● Hard
Calculation amount   ○ Small   ● Normal   ○ Large
1) $1\ \mu F$.
2) $1\ mF$.
3) $0.5\ mF$.
4) It is impossible to adjust the power factor of the bus at one.

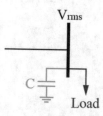

Fig. 1.13 The power system of problem 1.22

1.23. In the single-phase power system of Fig. 1.14, three loads have been connected to the power bus in parallel. Determine the capacitance of the shunt capacitor that needs to be connected to the bus to adjust its power factor at one for the following specifications. Moreover, calculate the current of the line after connecting the shunt capacitor to the bus:

$$Load\ 1 : (8 - j16)\ \Omega$$

$$Load\ 2 : (0.8 + j5.6)\ \Omega$$

$$Load\ 3 : S = 5\ kVA, \cos(\theta) = 0.8\ Lagging$$

$$V_{rms} = 200\ V, f = 60\ Hz$$

Difficulty level   ○ Easy   ○ Normal   ● Hard
Calculation amount   ○ Small   ● Normal   ○ Large
1) $100\ \mu F$, $20\ A$
2) $55\ \mu F$, $25\ A$
3) $800\ \mu F$, $25\ A$
4) $530\ \mu F$, $30\ A$

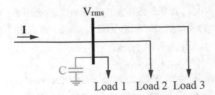

Fig. 1.14 The power system of problem 1.23

1.24. In the power system of Fig. 1.15, determine the reactive power of the shunt capacitor to keep the voltage of its bus at 1 p.u. In this problem, assume that $\cos(\sin^{-1}(0.1)) \equiv 0.995$.

Difficulty level   ○ Easy   ○ Normal   ● Hard
Calculation amount   ○ Small   ● Normal   ○ Large

1) 1.05 p. u.
2) 1.15 p. u.
3) 1.5 p. u.
4) 2.2 p. u.

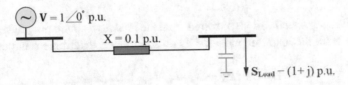

Fig. 1.15 The power system of problem 1.4

1.25. In the three-phase power system of Fig. 1.16, two balanced three-phase loads with the star and delta connections have been connected to a three-phase power supply. Calculate the line voltage of the loads for the following specifications:

$$E_{rms} = 4\,V, \mathbf{Z_1} = j2\,\Omega, \mathbf{Z_2} = (2 + j2)\,\Omega, \mathbf{Z_3} = j3\,\Omega, \mathbf{Z_4} = -j6\,\Omega$$

Difficulty level   ○ Easy   ○ Normal   ● Hard
Calculation amount   ○ Small   ● Normal   ○ Large

1) $2\sqrt{3}\,V$
2) $(6\angle 30°)\,V$
3) $6\sqrt{3}\,V$
4) $(6\sqrt{3}\angle 30°)\,V$

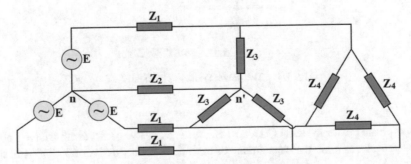

Fig. 1.16 The power system of problem 1.25

1.26. In the power system of Fig. 1.17, $\delta = 15^\circ$. If the value of δ increases and E_1 and E_2 are kept constant, which one of the following choices is correct? In this problem, assume that $\mathbf{I_{12}}$ always lags $\mathbf{E_2}$ and $\mathbf{Z} = jX$, $\mathbf{E_1} = E_1 < \delta$, $\mathbf{E_2} = E_2 < 0$.

Difficulty level   ○ Easy   ○ Normal   ● Hard

Calculation amount   ○ Small   ● Normal   ○ Large

1) $|\mathbf{I_{12}}|$ will increase and its phase angle with respect to $\mathbf{E_2}$ will increase.
2) $|\mathbf{I_{12}}|$ will decrease and its phase angle with respect to $\mathbf{E_2}$ will decrease.
3) $|\mathbf{I_{12}}|$ will increase and its phase angle with respect to $\mathbf{E_2}$ will decrease.
4) $|\mathbf{I_{12}}|$ will decrease and its phase angle with respect to $\mathbf{E_2}$ will increase.

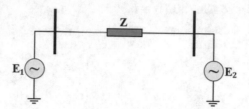

Fig. 1.17 The power system of problem 1.26

1.27. Three loads with the following specifications, resulted from the load flow simulation, have been connected to the power bus shown in Fig. 1.18. If all the loads are modeled by an admittance, determine it in per unit (p.u.):

$$\text{Load 1} : P_1 = 2\ p.u., PF = 0.8\ Lagging$$

$$\text{Load 2} : P_2 = 2\ p.u., PF = 0.8\ Leading$$

$$\text{Load 3} : P_3 = 2\ p.u., PF = 1$$

$$\mathbf{V} = (1\underline{/-12^\circ})\ V$$

Difficulty level   ○ Easy   ○ Normal   ● Hard

Calculation amount   ○ Small   ● Normal   ○ Large

1) $6\ p.u.$
2) $(2 - j)\ p.u.$
3) $(2 + j)\ p.u.$
4) $(2 - j2)\ p.u.$

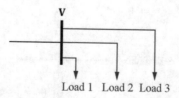

Load 1   Load 2   Load 3

Fig. 1.18 The power system of problem 1.27

1.28. At the end of a three-phase power system, 400 V, 50 Hz, three capacitor banks (with triangle configuration) have been connected to the system. Determine the capacitance of each bank if they deliver 600 kVAr to the system.

Difficulty level   ○ Easy   ○ Normal   ● Hard

Calculation amount   ○ Small   ● Normal   ○ Large

1) $5000\ \mu F$
2) $4000\ \mu F$
3) $0.004\ \mu F$
4) $0.005\ \mu F$

1.29. The single-line diagram of a balanced three-phase power system is shown in Fig. 1.19. In this problem $S_B = 100\ MVA$ and $V_B = 22\ kV$ in the first bus. Calculate the impedance seen from the first bus if the following specifications are given:

$$G : 22\ kV, 90\ MVA, X_G = 18\%$$

$$T_1 : 22/220\ kV, 50\ MVA, X_{T1} = 10\%$$

$$T_2 : 220/11\ kV, 40\ MW, X_{T2} = 6\%$$

$$T_3 : 22/110\ kV, 40\ MW, X_{T3} = 6.4\%$$

$$T_4 : 110/11\ kV, 40\ MW, X_{T4} = 8\%$$

$$M : 10.45\ kV, 66.5\ MVA, X_M = 18.5\%$$

$$TL_1 : 220\ kV, 48.4\ \Omega$$

$$TL_2 : 110\ kV, 65.5\ \Omega$$

Difficulty level   ○ Easy   ○ Normal   ● Hard
Calculation amount   ○ Small   ○ Normal   ● Large
1) $j0.14$
2) $j0.2$
3) $j0.22$
4) $j0.4$

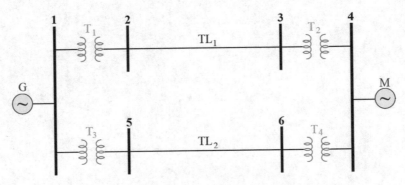

Fig. 1.19 The power system of problem 1.29

1.30. In the power system of Fig. 1.20, calculate the current of the load in per unit (p.u.) for the following specifications. In this problem, 100 V (in the generator side) and 1 kVA are chosen as the base voltage and power:

$$G : 100\ V$$

$$T_1 : 200/400\ V, 1\ kVA, X_{T1} = 0.1\ p.u.$$

$$\text{Line} : \mathbf{Z_{Line}} = j8\ \Omega$$

$$T_2 : 200/200\ V, 2\ kVA, X_{T2} = 0.1\ p.u.$$

$$\text{Load} : \mathbf{Z_{Load}} = j6\ \Omega$$

Difficulty level   ○ Easy   ○ Normal   ● Hard
Calculation amount   ○ Small   ○ Normal   ● Large

1) 0.25 $p.\,u.$
2) 1.5 $p.\,u.$
3) 0.5 $p.\,u.$
4) 1.25 $p.\,u.$

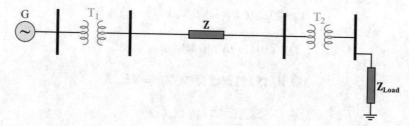

Fig. 1.20 The power system of problem 1.30

Abstract

In this chapter, the problems of the first chapter are fully solved, in detail, step by step, and with different methods.

2.1. As we know, $\cos(t)$ is usually chosen as the reference phasor. Hence, its phase angle is zero. Moreover, the amplitude of a phasor is normally shown in root-mean-square (rms) value. Therefore, the phasor representation of the signal of $\sqrt{2}\cos(t)$ can be calculated as follows. Herein, "$\angle$" is the symbol of phase angle.

$$v(t) = \sqrt{2}\cos(t) \Rightarrow \mathbf{V} = \frac{1}{\sqrt{2}} \times \left(\sqrt{2}\,\angle 0°\right) = 1\,\angle 0° \Rightarrow \mathbf{V} = 1\,V$$

Choice (1) is the answer.

2.2. The relation below holds between the signals of $\sin(t)$ and $\cos(t)$.

$$\sin(t) = \cos\left(t - 90°\right)$$

The signal of $\cos(t)$ is usually chosen as the reference phasor. In addition, the amplitude of a phasor is normally shown in root-mean-square (rms) value. Therefore, the phasor of the signal of $\sqrt{2}\sin(t)$ can be represented as follows.

$$i(t) = \sqrt{2}\sin(t) = \sqrt{2}\cos(t - 90°) \Rightarrow \mathbf{I} = \frac{1}{\sqrt{2}} \times \left(\sqrt{2}\,\angle{-90°}\right) \Rightarrow \mathbf{I} = (1\,\angle{-90°})\,A$$

Choice (4) is the answer.

2.3. The phasor of $\cos(2t + 30°)$ can be defined as follows.

$$\frac{1}{\sqrt{2}} \times (1\,\angle 30°) = \frac{1}{\sqrt{2}}\,\angle 30°$$

Herein, the signal of $\cos(t)$ is chosen as the reference phasor, and the amplitude of the phasor is presented in root-mean-square (rms) value. Choice (4) is the answer.

2.4. As we know, the relation below exists between the signals of $\sin(t)$ and $\cos(t)$.

$$\sin(t) = \cos(t - 90°) \Rightarrow \sin(t - 60°) = \cos(t - 150°)$$

Therefore, the phasor of $10 \sin(t - 60°)$ can be represented as follows.

$$\frac{1}{\sqrt{2}} \times (10\,\underline{/-150°}) = \left(5\sqrt{2}\,\underline{/-150°}\right)$$

Herein, the signal of $\cos(t)$ is chosen as the reference phasor, and the amplitude of the phasor is presented in root-mean-square (rms) value. Choice (3) is the answer.

2.5. Based on the information given in the problem, we have the following specifications:

$$v(t) = 110 \cos(\omega t + 30°)\ V \tag{1}$$

$$i(t) = 0.5 \cos(\omega t - 30°)\ A \tag{2}$$

Transferring to phasor domain:

$$\mathbf{V} = \left(55\sqrt{2}\,\underline{/30°}\right)V \tag{3}$$

$$\mathbf{I} = \left(\frac{\sqrt{2}}{4}\,\underline{/-30°}\right)A \tag{4}$$

The impedance is defined as follows:

$$\mathbf{Z} = \frac{\mathbf{V}}{\mathbf{I}} \tag{5}$$

Solving (3)–(5):

$$\mathbf{Z} = \frac{55\sqrt{2}\,\underline{/30°}}{\frac{\sqrt{2}}{4}\,\underline{/-30°}} \Rightarrow \mathbf{Z} = (220\,\underline{/60°})\ \Omega = \left(110 + j110\sqrt{3}\right)\Omega$$

The real and imaginary parts of impedance are called resistance and reactance, respectively. Thus:

$$R = Real\{\mathbf{Z}\} = 110\ \Omega$$

$$X = Imag\{\mathbf{Z}\} = 110\sqrt{3}\ \Omega$$

Choice (2) is the answer.

Fig. 2.1 The power system of solution of problem 2.5

2.6. Based on the information given in the problem, we have the following specifications:

$$v(t) = 100\sqrt{2}\cos(t) \ V \tag{1}$$

$$i(t) = \sqrt{2}\cos(t - 30°) \ A \tag{2}$$

Transferring to phasor domain:

$$\mathbf{V} = (100\angle 0°) \ V = 100 \ V \tag{3}$$

$$\mathbf{I} = (1\angle-30°) \ A \tag{4}$$

The admittance is defined as follows:

$$\mathbf{Y} = \frac{\mathbf{I}}{\mathbf{V}} \tag{5}$$

Solving (3)–(5):

$$\mathbf{Y} = \frac{1\angle-30°}{100} \Rightarrow \mathbf{Y} = (0.01\angle-30°) \ mho = (0.005\sqrt{3} - j0.005) \ mho$$

The real and imaginary parts of admittance are called conductance and susceptance, respectively.

$$G = Real\{\mathbf{Y}\} = 0.005\sqrt{3} \ mho$$

$$B = Imag\{\mathbf{Y}\} = 0.005 \ mho$$

Choice (3) is the answer.

2.7. Based on the information given in the problem, we have the following specifications:

$$V = 20 \ kV, S = 200 \ MVA, \mathbf{Z} = j0.2 \ p.u. \tag{1}$$

$$V_B = 21 \ kV, S_B = 100 \ MVA \tag{2}$$

The impedance of the generator has been presented in per unit (p.u.) value based on its rated quantities. Now, we need to update its per unit value based on the new base MVA and voltage as follows:

$$\mathbf{Z}_{new,p.u.} = \mathbf{Z}_{old,p.u.} \times \frac{S_{B,new}}{S_{B,old}} \times \left(\frac{V_{B,old}}{V_{B,new}}\right)^2 = j0.2 \times \frac{100 \ MVA}{200 \ MVA} \times \left(\frac{20 \ kV}{21 \ kV}\right)^2 = j0.0907 \tag{3}$$

$$\Rightarrow \mathbf{Z}_{new,percent} = \mathbf{Z}_{new,p.u.} \times 100 \Rightarrow \mathbf{Z}_{new,percent} = j9.07\%$$

Choice (4) is the answer.

2.8. Based on the information given in the problem, we have the following specifications:

$$V = 14 \ kV, S = 500 \ MVA, X = 1.1 \ p.u. \tag{1}$$

$$V_B = 20 \ kV, S_B = 100 \ MVA \tag{2}$$

The impedance of the generator has been presented in per unit (p.u.) based on its rated quantities. We need to update its per unit value based on the new base MVA and voltage as follows:

$$X_{new,p.u.} = X_{old,p.u.} \times \frac{S_{B,new}}{S_{B,old}} \times \left(\frac{V_{B,old}}{V_{B,new}}\right)^2 = 1.1 \times \frac{100 \; MVA}{500 \; MVA} \times \left(\frac{14 \; kV}{20 \; kV}\right)^2 = 0.1078 \tag{3}$$

$$\Rightarrow X_{new,percent} = X_{new,p.u.} \times 100 \Rightarrow X_{new,percent} = 10.78\%$$

Choice (2) is the answer.

2.9. For this problem, we only need to apply KCL in the bus as follows. KCL can be applied in this bus, since all the currents have the same angular frequency ($\omega = 10 \; rad/sec$):

$$-i_1(t) + i_2(t) + i_3(t) + i_4(t) = 0 \tag{1}$$

Based on the information given in the problem, we know that:

$$i_1(t) = 10\cos(10t) \; A \tag{2}$$

$$i_2(t) = 10\sin(10t) \; A \tag{3}$$

$$i_4(t) = 10\sqrt{2}\cos(10t + 45^\circ) \; A \tag{4}$$

It is better to represent the currents in phasor domain, as can be seen in the following. Herein, the signal of $\cos(t)$ is chosen as the reference phasor, the amplitude of the phasor is presented in root-mean-square (rms) value, and "$\angle\underline{\hspace{0.5em}}$" is the symbol of phase angle:

$$-\mathbf{I_1} + \mathbf{I_2} + \mathbf{I_3} + \mathbf{I_4} = 0 \tag{5}$$

$$\mathbf{I_1} = \left(5\sqrt{2} \angle 0^\circ\right) A = 5\sqrt{2} \; A \tag{6}$$

$$\mathbf{I_2} = \left(5\sqrt{2} \angle -90^\circ\right) A \tag{7}$$

$$\mathbf{I_4} = \left(10 \angle 45^\circ\right) A \tag{8}$$

Solving (5)–(8):

$$-5\sqrt{2} + \left(5\sqrt{2}\angle -90^\circ\right) + \mathbf{I_3} + \left(10\angle 45^\circ\right) = 0 \Rightarrow \mathbf{I_3}$$
$$= 5\sqrt{2} - \left(5\sqrt{2}\angle -90^\circ\right) - \left(10\angle 45^\circ\right) \Rightarrow \mathbf{I_3} = 0 \; A \tag{8}$$

By transferring back to time domain, we can write:

$$i_3(t) = 0 \; A$$

Choice (4) is the answer.

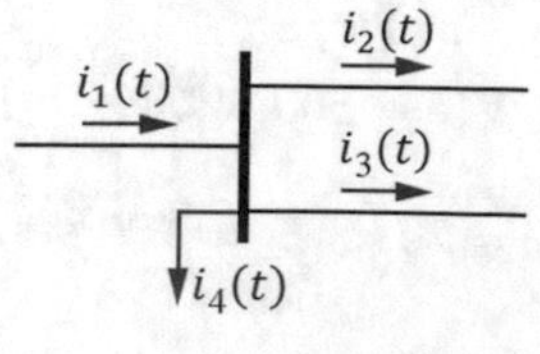

Fig. 2.2 The power system of solution of problem 2.9

2.10. As we know, active power is consumed by resistance of load and can be calculated for a single-phase system as follows:

$$P = R(I_{rms})^2 \tag{1}$$

Based on the information given in the problem, we have the following specifications:

$$V_{rms} = 200 \, V \tag{2}$$

$$\mathbf{Z_1} = (8 - j6) \, \Omega \tag{3}$$

$$\mathbf{Z_2} = (3 + j4) \, \Omega \tag{4}$$

Using Ohm's law for the first load:

$$I_{rms,1} = \frac{V_{rms}}{|\mathbf{Z_1}|} = \frac{200}{|8 - j6|} = 20 \, A \tag{5}$$

Applying Ohm's law for the second load:

$$I_{rms,2} = \frac{V_{rms}}{|\mathbf{Z_2}|} = \frac{200}{|3 + j4|} = 40 \, A \tag{6}$$

Solving (1), (3), and (5):

$$P_1 = R_1(I_{rms,1})^2 = 8 \times 20^2 = 3200 \, W \tag{7}$$

Solving (1), (4), and (6):

$$P_2 = R_2(I_{rms,2})^2 = 3 \times 40^2 = 4800 \, W \tag{8}$$

Therefore:

$$P_{Total} = P_1 + P_2 = 3200 + 4800 = 8000 \, W \Rightarrow P_{Total} = 8 \, kW$$

Choice (1) is the answer.

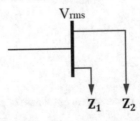

Fig. 2.3 The power system of solution of problem 2.10

2.11. Instantaneous power of a single-phase power system can be calculated as follows:

$$p(t) = v(t)i(t) \tag{1}$$

Based on the information given in the problem, we have the following specifications:

$$v(t) = 110\sqrt{2}\cos(120\pi t) \, V \tag{2}$$

$$i(t) = 2\sqrt{2}\cos\left(120\pi t - 60^\circ\right) A \qquad (3)$$

Solving (1)–(3):

$$p(t) = 110\sqrt{2}\cos\left(120\pi t\right) \times 2\sqrt{2}\cos\left(120\pi t - 60^\circ\right) = 440\cos\left(120\pi t\right)\cos\left(120\pi t - 60^\circ\right) \qquad (4)$$

From trigonometry, we know that:

$$\cos\left(a\right)\cos\left(b\right) = \frac{1}{2}\left(\cos\left(a+b\right) + \cos\left(a-b\right)\right) \qquad (5)$$

Solving (4) and (5):

$$p(t) = 220\left(\cos\left(240\pi t - 60^\circ\right) + \cos\left(60^\circ\right)\right) \Rightarrow p(t) = 110 + 220\cos\left(240\pi t - 60^\circ\right) W$$

Choice (4) is the answer.

2.12. Based on the information given in the problem, we have the following specifications:

$$\mathbf{V_1} = (10\underline{/30^\circ})\,V, \mathbf{V_2} = \left(5\sqrt{3}\underline{/0^\circ}\right)V, \mathbf{Z} = j5\,\Omega \qquad (1)$$

The current flowing from bus 1 to bus 2 can be calculated as follows:

$$\mathbf{I_{12}} = \frac{\mathbf{V_1} - \mathbf{V_2}}{\mathbf{Z}} = \frac{(10\underline{/30^\circ}) - \left(5\sqrt{3}\underline{/0^\circ}\right)}{j5} = \frac{5\sqrt{3} + j5 - 5\sqrt{3}}{j5} = 1\,A \qquad (2)$$

The complex power transferred from bus 1 to bus 2 can be calculated as follows:

$$\mathbf{S_{12}} = \mathbf{V_1}\mathbf{I_{12}^*} = (10\underline{/30^\circ})(1)^* = (10\underline{/30^\circ})\,VA = \left(5\sqrt{3} + j5\right)VA \Rightarrow \begin{cases} P_{12} = 5\sqrt{3}\ W \\ Q_{12} = 5\ VAr \end{cases}$$

Choice (4) is the answer.

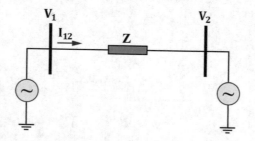

Fig. 2.4 The power system of solution of problem 2.12

2.13. Based on the information given in the problem, we have the following specifications:

$$\mathbf{E_1} = 200\ \underline{/-30^\circ}\,V, \mathbf{E_2} = 200\underline{/0^\circ}\,V, \mathbf{Z} = j5\,\Omega \qquad (1)$$

The active and reactive powers flowing in the transmission line from bus 1 to bus 2 can be calculated as follows:

$$P_{12} = \frac{|V_1||V_2|}{X}\sin\left(\theta_1 - \theta_2\right) = \frac{200 \times 200}{5}\sin\left(-30 - 0\right) = -4000 < 0\ W \qquad (2)$$

$$P_{21} = \frac{|V_1||V_2|}{X} \sin(\theta_2 - \theta_1) = \frac{200 \times 200}{5} \sin(0 - (-30)) = 4000 > 0 \ W \qquad (3)$$

$$Q_{12} = \frac{|V_1|}{X}(|V_1| - |V_2|\cos(\theta_1 - \theta_2)) = \frac{200}{5}(200 - 200\cos(-30 - 0)) \approx 1071 \ VAr > 0 \qquad (4)$$

$$Q_{21} = \frac{|V_2|}{X}(|V_2| - |V_1|\cos(\theta_2 - \theta_1)) = \frac{200}{5}(200 - 200\cos(0 - (-30))) \approx 1071 \ VAr > 0 \qquad (5)$$

As can be noticed from (2) and (3), $P_{12} < 0$ and $P_{21} > 0$. Therefore, the active power flows from bus 2 to bus 1. In other words, the first and the second electric machines are working as a motor and a generator, respectively.

However, as can be noticed from (4) and (5), $Q_{12} = Q_{21} > 0$. Thus, the reactive power is generated by both machines and ultimately consumed in the transmission line.

Choice (4) is the answer.

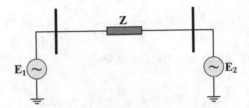

Fig. 2.5 The power system of solution of problem 2.13

2.14. Based on the information given in the problem, we have the following specifications:

$$V_R = 20 \ kV, S_R = 100 \ MVA \qquad (1)$$

$$V_R = 20 \ kV, Q_R = 200 \ MVAr \qquad (2)$$

The reactance of the reactor can be calculated as follows:

$$Q_R = \frac{(V_R)^2}{X_R} \Rightarrow X_R = \frac{(V_R)^2}{Q_R} = \frac{(20 \ kV)^2}{200 \ MVAr} = 2 \ \Omega \qquad (3)$$

The base impedance in the bus can be calculated as follows:

$$S_B = \frac{(V_B)^2}{Z_B} \Rightarrow Z_B = \frac{(V_B)^2}{S_B} = \frac{(20 \ kV)^2}{100 \ MVA} = 4 \ \Omega \qquad (4)$$

The reactance of the reactor in per unit (p.u.) can be determined by using (3) and (4):

$$X_{R,p.u.} = \frac{X_R}{Z_B} \Rightarrow X_{R,p.u.} = \frac{2}{4} = 0.5 \ \Omega$$

Choice (2) is the answer.

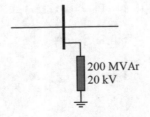

200 MVAr
20 kV

Fig. 2.6 The power system of solution of problem 2.14

2.15. Figure 2.7 shows the single-line diagram of the power system with the indicated zones. Based on the information given in the problem, we have the following specifications:

$$G : 20\,kV, 300\,MVA \tag{1}$$

$$T_1 : 20/200\,kV, 375\,MVA \tag{2}$$

$$T_2 : 180/9\,kV, 300\,MW \tag{3}$$

$$Load : 9\,kV, 180\,MW \tag{4}$$

$$V_{B1} = 20\,kV, S_B = 300\,MVA \tag{5}$$

The resistance of the purely resistive load can be calculated as follows:

$$P = \frac{(V)^2}{R} \Rightarrow R = \frac{(V)^2}{P} = \frac{(9\,kV)^2}{180\,MW} = 0.45\,\Omega \tag{6}$$

The base voltage in the third zone can be calculated as follows:

$$V_{B3} = 20\,kV \times \frac{200}{20} \times \frac{9}{180} = 10\,kV \tag{7}$$

The base impedance in the third zone can be calculated as follows:

$$S_B = \frac{(V_{B3})^2}{Z_B} \Rightarrow Z_{B3} = \frac{(V_{B3})^2}{S_B} = \frac{(10\,kV)^2}{300\,MVA} = \frac{1}{3}\,\Omega \tag{8}$$

The resistance of the load in per unit (p.u.) can be determined by using (6) and (8):

$$R_{p.u.} = \frac{R}{Z_B} \Rightarrow R_{p.u.} = \frac{0.45}{\frac{1}{3}} \Rightarrow R_{p.u.} = 1.35\,p.u.$$

Choice (2) is the answer.

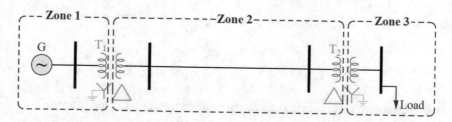

Fig. 2.7 The power system of solution of problem 2.15

2.16. Based on the information given in the problem, we have the following specifications:

$$\sin\left(15^\circ\right) \equiv 0.25, \cos\left(15^\circ\right) \equiv 0.96 \tag{1}$$

$$\mathbf{V_1} = (1\angle 0^\circ)\,p.u. \tag{2}$$

$$\mathbf{V_2} = (1\angle -15^\circ)\,p.u. \tag{3}$$

$$\mathbf{Z} = j0.5 \ p.u. \tag{4}$$

The active power flowing through the transmission line can be calculated as follows:

$$P_{12} = \frac{|V_1||V_2|}{X} \sin(\theta_1 - \theta_2) \Rightarrow P_{12} = \frac{1 \times 1}{0.5} \sin(0 - (-15°)) = 2 \sin(15°) \tag{5}$$

Solving (1) and (5):

$$P_{12} = 2 \times 0.25 \Rightarrow P_{12} = 0.5 \ p.u \tag{6}$$

The reactive power flowing through the transmission line can be calculated as follows:

$$Q_{21} = \frac{|V_2|}{X}(|V_2| - |V_1| \cos(\theta_2 - \theta_1)) = \frac{1}{0.5}(1 - \cos(-15° - 0)) = 2(1 - \cos(15°)) \tag{7}$$

Solving (1) and (7):

$$Q_{21} = 2(1 - 0.96) = 0.08 \ p.u \tag{8}$$

In bus 2, we can write:

$$Q_{21} + Q = 0 \Rightarrow Q = -Q_{21} \Rightarrow Q = -0.08 \ p.u.$$

Choice (4) is the answer.

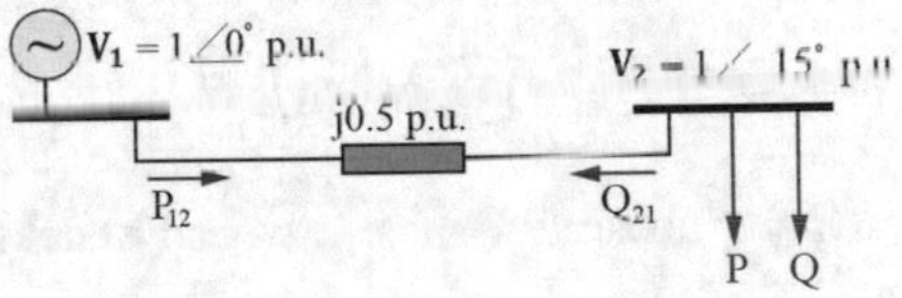

Fig. 2.8 The power system of solution of problem 2.16

2.17. Based on the information given in the problem, we have the following specifications:

$$\text{Inductive Load}: P_1 = 60 \ kW, Q_1 = 660 \ kVAr \tag{1}$$

$$\text{Capacitive Load}: P_2 = 240 \ kW, PF_2 = 0.8 \tag{2}$$

From (1), we can write:

$$\mathbf{S_1} = P_1 + jQ_1 = (60 + j660) \ kVA \tag{3}$$

In addition, from (2), we can write:

$$\mathbf{S_2} = \frac{P_2}{\cos(PF_2)} < \cos^{-1}(PF_2) \Rightarrow \mathbf{S_2} = \frac{240 \ kW}{0.8} < -\cos^{-1}(0.8) = (300 \underline{/-36.9°}) \ kVA$$
$$= (240 - j180) \ kVA \tag{4}$$

In (4), a negative sign was applied in phase angle for the complex power, as the power factor of the load is capacitive:

$$\mathbf{S_{Total}} = \mathbf{S_1} + \mathbf{S_2} = (60 + j660)\ kVA + (240 - j180)\ kVA \Rightarrow \mathbf{S_{Total}} = (300 - j480)\ kVA$$

Choice (2) is the answer.

2.18. Based on the information given in the problem, we have the following specifications:

$$G : 20\ kV, 300\ MVA, X_G = 20\% \tag{1}$$

$$T_1 : 20/230\ kV, 150\ MVA, X_T = 0.1\ p.u. \tag{2}$$

$$Line : 176.33\ km, X_{Line} = 1\ \Omega/km \tag{3}$$

$$V_{B1} = 20\ kV, S_B = 300\ MVA \tag{4}$$

The impedance of the generator will not change as its rated values have been chosen as the base quantities. Hence:

$$X_{G,p.u.} = 0.2\ (5) \tag{5}$$

Now, we need to update the per unit (p.u.) value of the transformer's impedance based on the new base MVA and voltage as follows:

$$X_{new,p.u.} = X_{old,p.u.} \times \frac{S_{B,new}}{S_{B,old}} \times \left(\frac{V_{B,old}}{V_{B,new}}\right)^2 \tag{6}$$

$$X_{T,new,p.u.} = 0.1 \times \frac{300}{150} \times \left(\frac{20}{20}\right)^2 = 0.2\ p.u. \tag{7}$$

To present the impedance of the line in per unit (p.u.) value, we need to determine the base impedance in the second zone (see Fig. 2.9.2), as follows:

$$V_{B2} = 20\ kV \times \frac{230}{20} = 230\ kV \tag{8}$$

$$S_B = \frac{(V_{B2})^2}{Z_B} \Rightarrow Z_{B2} = \frac{(V_{B2})^2}{S_B} = \frac{(230\ kV)^2}{300\ MVA} = 176.33\ \Omega \tag{9}$$

$$\mathbf{Z_{Line,p.u.}} = \frac{\mathbf{Z_{Line}}}{Z_{B2}} \Rightarrow \mathbf{Z_{Line,p.u.}} = \frac{176.33\ km \times 1\ \Omega/km}{176.33} = 1\ p.u. \tag{10}$$

Figure 2.9.3 shows the impedance diagram of the power system by using (5), (7), and (10). The Thevenin reactance, seen from the end of the transmission line, can be calculated as follows:

$$X_{Thevenin,p.u.} = X_{G,p.u.} + X_{T,new,p.u.} + X_{Line,p.u.} = 1 + 0.2 + 0.2 \Rightarrow X_{Thevenin,p.u.} = 1.4\ p.u.$$

Choice (4) is the answer.

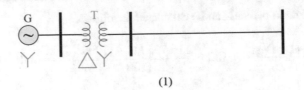

(1)

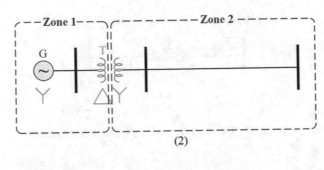

(2)

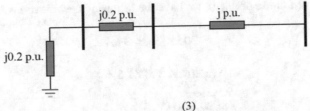

(3)

Fig. 2.9 The power system of solution of problem 2.18

2.19. Based on the information given in the problem, we have the following specifications:

$$\text{Line}: \mathbf{Z} = (10 + j40) \; \Omega/phase \tag{1}$$

$$\text{Load}: V = 100 \; kV, S = 50 \; MVA, PF = 0.8 \; Lagging \tag{2}$$

First, we should solve the problem for the single-phase system. The voltage of the load is chosen as the reference. Hence:

$$\mathbf{V_{2,ph}} = \left(\frac{100}{\sqrt{3}} < 0 \right) kV \tag{3}$$

The current of the load can be calculated as follows:

$$\mathbf{I_{Load}} = \frac{S_{Load}}{\sqrt{3}V_{Load}} < \cos^{-1}(0.8) = \frac{50 \; MVA}{\sqrt{3} \times 100 \; kV} \underline{/37°} = (288.6 \underline{/36.9°}) \; A \tag{3}$$

Applying KVL:

$$\mathbf{V_{1,ph}} = \mathbf{ZI} + \mathbf{V_{2,ph}} = (10 + j40) \times (288.6 \underline{/36.9°}) + \left(\frac{100}{\sqrt{3}} < 0 \right) = (66.972 + j7.505) \; kV \tag{4}$$

$$= (67.391 \underline{/6.4°}) \; kV$$

$$\left| \mathbf{V_{1,L}} \right| = \sqrt{3} \left| \mathbf{V_{1,ph}} \right| = 116.725 \; kV \tag{5}$$

The voltage drop of the line in percent can be calculated as follows:

$$\Delta V\% = \frac{|\mathbf{V_{1,L}}| - |\mathbf{V_{2,L}}|}{|\mathbf{V_{2,L}}|} \times 100 = \frac{116.725\ kV - (100 < 0)\ kV}{(100 < 0)\ kV} \times 100 \Rightarrow \Delta V\% = 16\%$$

Choice (2) is the answer.

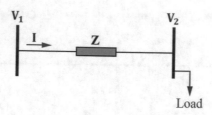

Fig. 2.10 The power system of solution of problem 2.19

2.20. Based on the information given in the problem, we have the following specifications:

$$V_{B1} = 20\ kV, S_B = 3\ MVA \tag{1}$$

$$G : 20\ kV, 3\ MVA, 3\% \tag{2}$$

$$T_1 : 20/230\ kV, 3\ MVA, 5\% \tag{3}$$

$$T_2 : 230/11\ kV, 3\ MVA, 5\% \tag{4}$$

$$Load : 11\ kV, 0.2\ MVA, 0.8\ Lagging \tag{5}$$

$$M : 11\ kV, 1\ MVA, 5\% \tag{6}$$

$$C : 0.5\ MVA \tag{7}$$

The base voltage in the third zone can be calculated as follows:

$$V_{B3} = 20\ kV \times \frac{230}{20} \times \frac{11}{230} = 11\ kV \tag{8}$$

The impedance of the load can be calculated as follows:

$$\mathbf{Z_{Load}} = \frac{(V_{Load})^2}{S_{Load}} < \cos^{-1}(PF_{Load}) \Rightarrow \mathbf{Z_{Load}} = \frac{(11\ kV)^2}{0.2\ MVA} \angle 36.9° \tag{9}$$

The base impedance in the third zone can be calculated as follows:

$$S_B = \frac{(V_{B3})^2}{Z_{B3}} \Rightarrow Z_{B3} = \frac{(V_{B3})^2}{S_B} = \frac{(11\ kV)^2}{3\ MVA} = \frac{121}{3}\ \Omega \tag{10}$$

The impedance of the load in per unit (p.u.) can be determined by using (9) and (10):

$$\mathbf{Z_{Load,p.u.}} = \frac{\mathbf{Z_{Load}}}{Z_{B3}} \Rightarrow \mathbf{Z_{Load,p.u.}} = \frac{\dfrac{(11\ kV)^2}{0.2\ MVA} \angle 36.9°}{\dfrac{121}{3}} \Rightarrow \mathbf{Z_{Load,p.u.}} = (15 \angle 36.9°)\ p.u.$$

$$= (12 + j9)\ p.u.$$

Choice (1) is the answer.

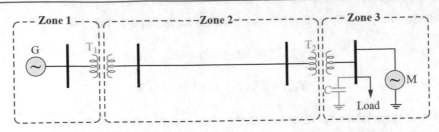

Fig. 2.11 The power system of solution of problem 2.20

2.21. The power factor of the bus can be calculated as follows:

$$PF_{Total} = \frac{P_{Total}}{S_{Total}} = \frac{\sum P_i}{\sqrt{\left(\sum P_i\right)^2 + \left(\sum Q_i\right)^2}} \tag{1}$$

Based on the information given in the problem, we have the following specifications:

$$\text{Load 1}: P_1 = 25\ kW, Q_1 = 25\ kVAr \tag{2}$$

$$\text{Load 2}: S_2 = 15\ kVA,\ \cos(\theta_2) = 0.8\ Leading \tag{3}$$

$$\text{Load 3}: P_3 = 11\ kW,\ \cos(\theta_3) = 1 \tag{4}$$

From (3), we can write:

$$P_2 = S_2 \cos(\theta_2) \Rightarrow P_2 = 15 \times 0.8 = 12\ kW \tag{5}$$

$$Q_2 = -S_2 \sin(\theta_2) \Rightarrow Q_2 = -15 \times \sqrt{1 - (0.8)^2} = -9\ kVAr \tag{6}$$

In (6), a negative sign was applied in the formula, as the power factor of the load is leading.

From (4), we can conclude the following term, since the power factor is unit:

$$Q_3 = 0 \tag{7}$$

Solving (1)–(7):

$$PF_{Total} = \frac{25 + 12 + 11}{\sqrt{(25 + 12 + 11)^2 + (25 + (-9) + 0)^2}} = \frac{48}{\sqrt{(48)^2 + (16)^2}} \Rightarrow PF_{Total} = 0.94$$

Since $\sum Q_i = 16\ kVAr > 0$, the total power factor is lagging. Choice (1) is the answer.

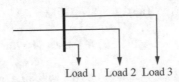

Load 1   Load 2 Load 3

Fig. 2.12 The power system of solution of problem 2.21

2.22. Based on the information given in the problem, we have the following specifications:

$$Load : S = 20\ kVA,\ \cos(\theta) = 0.8\ Lagging \tag{1}$$

$$V_{rms} = 200\ V, f = 50\ Hz, \pi \cong 3 \tag{2}$$

Since the final power factor of the bus must be unit, the whole reactive power of the load must be supplied by the shunt capacitor. In other words, the net reactive power of the bus must be zero:

$$Q_{Net} = Q_{Load} + (-Q_C) = 0 \Rightarrow Q_C = Q_{Load} \tag{3}$$

From (1), we can write:

$$Q_{Load} = S_{Load}\sin(\theta_{Load}) = 20 \times \sqrt{1 - (0.8)^2} = 12\ kVAr \tag{4}$$

Solving (3) and (4):

$$Q_C = 12\ kVAr \tag{5}$$

As we know, the reactive power of a single-phase capacitor can be determined as follows:

$$Q_C = \frac{V_{rms}^{\ 2}}{X_c} = \omega C V_{rms}^{\ 2} = 2\pi f C V_{rms}^{\ 2} \tag{6}$$

Solving (2), (5), and (6):

$$12000 = 2 \times 3 \times 50 \times C \times 200^2 \Rightarrow C = \frac{12000}{12 \times 10^6} \Rightarrow C = 1\ mF$$

Choice (2) is the answer.

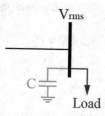

Fig. 2.13 The power system of solution of problem 2.22

2.23. Based on the information given in the problem, we have the following specifications:

$$Load\ 1 : (8 - j16)\ \Omega \tag{1}$$

$$Load\ 2 : (0.8 + j5.6)\ \Omega \tag{2}$$

$$Load\ 3 : S = 5\ kVA,\ \cos(\theta) = 0.8\ Lagging \tag{3}$$

$$V_{rms} = 200\ V, f = 60\ Hz \tag{4}$$

The total reactive power of the loads must be supplied by the shunt capacitor, since the final power factor of the bus is adjusted at one. In other words, the net reactive power of the bus must be zero:

$$Q_{Net} = Q_1 + Q_2 + Q_3 + (-Q_C) = 0 \Rightarrow Q_C = Q_1 + Q_2 + Q_3 \tag{5}$$

From (1) and (4), we can write:

$$\mathbf{S_1} = \frac{(V_{rms})^2}{\mathbf{Z_1^*}} = \frac{200^2}{(8 - j16)^*} = (1 + j7) \, kVA \tag{6}$$

From (2) and (4), we have:

$$\mathbf{S_2} = \frac{(V_{rms})^2}{\mathbf{Z_2^*}} = \frac{200^2}{(0.8 + j5.6)^*} = (1 - j2) \, kVA \tag{7}$$

From (3), we can write:

$$\mathbf{S_3} = S_3 < \cos^{-1}(PF_3) = 5 < \cos^{-1}(0.8) \, kVA = (5\underline{/36.8°}) \, kVA = (4 + j3) \, kVA \tag{8}$$

Solving (5)–(8):

$$Q_C = 7 + (-2) + 3 = 8 \, kVAr \tag{9}$$

As we know, reactive power of a single-phase capacitor can be determined as follows:

$$Q_C = \frac{V_{rms}^2}{X_c} = \omega C V_{rms}^2 = 2\pi f C V_{rms}^2 \tag{10}$$

Solving (4), (9), and (10):

$$8000 = 2 \times 3.14 \times 60 \times C \times 200^2 \Rightarrow C = \frac{8000}{15.072 \times 10^6} \Rightarrow C = 530 \, \mu F$$

As it was mentioned earlier, after connecting the shunt capacitor to the power bus, the net reactive power of the bus is zero because its power factor is unit. The current of the line can be calculated as follows:

$$\mathbf{S} = \mathbf{VI^*} \Rightarrow \mathbf{I} = \left(\frac{\mathbf{S}}{\mathbf{V}}\right)^* = \left(\frac{(P_1 + P_2 + P_3) + 0}{V_{rms} \underline{/0°}}\right)^* = \left(\frac{(1000 + 1000 + 4000) + 0}{200 < 0}\right)^*$$

$$\Rightarrow \mathbf{I} = 30 \, A$$

Choice (4) is the answer.

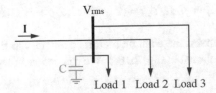

Fig. 2.14 The power system of solution of problem 2.23

2.24. Based on the information given in the problem, we have the following specifications:

$$|V_1| = |V_2| = 1 \, p.u. \tag{1}$$

$$\theta_1 = 0 \tag{2}$$

$$X = 0.1 \, p.u. \tag{3}$$

$$P_{12} = 1 \, p.u. \tag{4}$$

$$\cos\left(\sin^{-1}(0.1)\right) \equiv 0.995 \tag{5}$$

The active power flowing through the transmission line can be calculated as follows:

$$P_{12} = \frac{|V_1||V_2|}{X}\sin(\theta_1 - \theta_2) \Rightarrow 1 = \frac{1 \times 1}{0.1}\sin(0 - \theta_2) \Rightarrow \sin(\theta_2) = -0.1 \Rightarrow \theta_2 = \sin^{-1}(-0.1) \,(6) \tag{6}$$

The reactive power flowing through the transmission line can be calculated as follows:

$$Q_{21} = \frac{|V_2|}{X}\left(|V_2| - |V_1|\cos(\theta_2 - \theta_1)\right) = \frac{1}{0.1}(1 - \cos(\theta_2 - 0)) = 10(1 - \cos(\theta_2)) \tag{7}$$

Solving (5)–(7):

$$Q_{21} = 10\left(1 - \cos\left(\sin^{-1}(-0.1)\right)\right) = 10\left(1 - \cos\left(-\sin^{-1}(0.1)\right)\right) = 10\left(1 - \cos\left(\sin^{-1}(0.1)\right)\right) = 10(1 - 0.995)$$
$$= 0.05 \, p.u. \tag{8}$$

To keep the voltage of the bus at 1 p.u., the whole reactive power of the bus must be compensated by the shunt capacitor. In other words, the net reactive power of the bus must be zero:

$$Q_{Net} = Q_{21} + (-Q_C) + Q_{Load} = 0 \Rightarrow Q_C = Q_{21} + Q_{Load} = 0.05 + 1 \Rightarrow Q_C = 1.05 \, p.u.$$

Choice (1) is the answer.

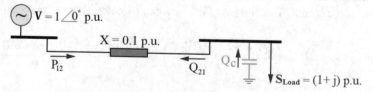

Fig. 2.15 The power system of solution of problem 2.24

2.25. Based on the information given in the problem, we have the following specifications:

$$E_{rms} = 4 \, V, \, \mathbf{Z_1} = j2 \, \Omega, \, \mathbf{Z_2} = (2 + j2) \, \Omega, \, \mathbf{Z_3} = j3 \, \Omega, \, \mathbf{Z_4} = -j6 \, \Omega \tag{1}$$

To solve this problem, we should convert the triangle (delta) connection to the star (wye) connection (see Fig. 2.16.2) and analyze the single-phase system shown in Fig. 2.16.3. We can connect the neutral node of the loads to each other, as the system is a balanced system.

As we know, the relation below exists between the impedance of a balanced triangle (delta) connection and its equivalent balanced star (wye) connection:

$$Z_Y = \frac{1}{3} Z_\Delta \Rightarrow Z_4' = \frac{1}{3} Z_4 = -j2 \ \Omega \tag{2}$$

Since the power system is a balanced system, no current flows through the neutral line and Z_2. Hence, no voltage drop occurs across Z_2. Therefore, we can ignore this impedance in the diagram of the single-phase system, as is illustrated in Fig. 2.16.3.

Applying voltage division rule:

$$V_{a'n'} = \frac{Z_3 \| Z_4'}{Z_1 + Z_3 \| Z_4'} \times E = \frac{(j3) \| (-j2)}{j2 + (j3) \| (-j2)} \times (4 \angle 0°) = \frac{\frac{6}{j}}{j2 + \frac{6}{j}} \times 4 = \frac{-j6}{-j4} \times 4 = 6 \ V \tag{3}$$

As we know, the relation below holds between the phase and line voltage:

$$V_{a'b'} = \sqrt{3} V_{a'n'} \angle 30° \tag{4}$$

$$\Rightarrow V_{a'b'} = \left(6\sqrt{3} \ \angle 30° \right) V$$

Choice (4) is the answer.

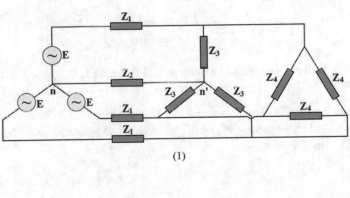

(1)

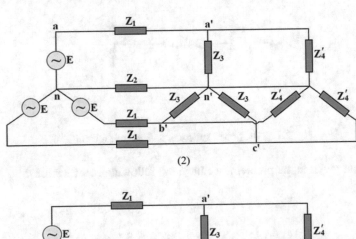

(2)

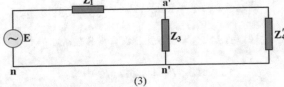

(3)

Fig. 2.16 The power system of solution of problem 2.25

2.26. Based on the information given in the problem, $\mathbf{I_{12}}$ always lags $\mathbf{E_2}$. Moreover, we have:

$$\delta = 15° \tag{1}$$

$$E_1 = Const., \quad E_2 = Const. \tag{2}$$

$$\mathbf{Z} = jX \tag{3}$$

$$\mathbf{E_1} = E_1 < \delta \tag{4}$$

$$\mathbf{E_2} = E_2 < 0 \tag{5}$$

The current in the transmission line can be calculated as follows:

$$\mathbf{I_{12}} = \frac{E_1 < \delta - E_2 < 0}{jX} = \frac{E_1 \cos(\delta) + jE_1 \sin(\delta) - E_2}{jX} = \frac{(E_1 \cos(\delta) - E_2) + jE_1 \sin(\delta)}{jX} \tag{6}$$

$$|\mathbf{I_{12}}| = \frac{\sqrt{(E_1 \cos(\delta) - E_2)^2 + (E_1 \sin(\delta))^2}}{X}$$

$$= \frac{\sqrt{E_1^2 \cos^2(\delta) - 2E_1 E_2 \cos(\delta) + E_2^2 + E_1^2 \sin^2(\delta)}}{X} = \frac{\sqrt{E_1^2 + E_2^2 - 2E_1 E_2 \cos(\delta)}}{X} \tag{7}$$

$$< \mathbf{I_{12}} = \tan^{-1}\left(\frac{E_1 \sin(\delta)}{E_1 \cos(\delta) - E_2}\right) - \frac{\pi}{2} \tag{8}$$

As can be noticed from (7), by increasing δ and keeping E_1 and E_2 constant, $\cos(\delta)$ will decrease, and consequently $|\mathbf{I_{12}}|$ will increase.

Moreover, as can be noticed from (8), by increasing δ and keeping E_1 and E_2 constant, $\sin(\delta)$ and $\cos(\delta)$ will increase and decrease, respectively, and consequently $<\mathbf{I_{12}}$ will increase (counter-clockwise). Therefore, the phase angle of the current with respect to $\mathbf{E_2}$ will decrease.

Choice (3) is the answer.

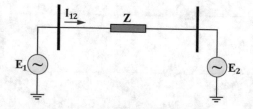

Fig. 2.17 The power system of solution of problem 2.26

2.27. Based on the information given in the problem, we have the following specifications:

$$Load\ 1 : P_1 = 2\ p.u., PF = 0.8\ Lagging \tag{1}$$

$$Load\ 2 : P_2 = 2\ p.u., PF = 0.8\ Leading \tag{2}$$

$$Load\ 3 : P_3 = 2\ p.u., PF = 1 \tag{3}$$

$$\mathbf{V} = (1 \underline{/-12°})\ V \tag{4}$$

The total active power of the loads can be calculated as follows:

$$P_{Total} = P_1 + P_2 + P_3 = 2 + 2 + 2 = 6 \, p.u. \tag{6}$$

The relation below holds between the active and reactive power of a load:

$$Q = P \tan(\theta) \tag{7}$$

The total reactive power of the loads can be calculated as follows:

$$Q_{Total} = Q_1 + Q_2 + Q_3 = P_1 \tan(\theta_1) + P_2 \tan(\theta_2) + P_3 \tan(\theta_3)$$

$$= P_1 \tan\left(\cos^{-1}(PF_1)\right) + P_2 \tan\left(\cos^{-1}(PF_2)\right) + P_3 \tan\left(\cos^{-1}(PF_3)\right)$$

$$= 2 \tan\left(\cos^{-1}(0.8)\right) + 2 \tan\left(-\cos^{-1}(0.8)\right) + 2 \tan\left(\cos^{-1}(1)\right)$$

$$= 2 \tan\left(\cos^{-1}(0.8)\right) - 2 \tan\left(\cos^{-1}(0.8)\right) + 2 \tan(0) = 0 \, p.u. \tag{8}$$

The total complex power is:

$$\mathbf{S_{Total}} = P_{Total} + Q_{Total} = 6 + j0 = 6 \, p.u. \tag{9}$$

The total current is:

$$\mathbf{I_{Total}} = \left(\frac{\mathbf{S_{Total}}}{\mathbf{V}}\right)^* = \left(\frac{6}{1\angle{-12°}}\right)^* = (6\angle{-12°}) \, p.u. \tag{10}$$

The equivalent admittance of the loads can be determined as follows:

$$\mathbf{Y} = \frac{\mathbf{I_{Total}}}{\mathbf{V}} = \frac{6\angle{-12°}}{1\angle{-12°}} \Rightarrow \mathbf{Y} = 6 \, p.u.$$

Choice (1) is the answer.

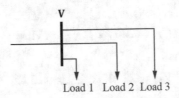

Fig. 2.18 The power system of solution of problem 2.27

2.28. Based on the information given in the problem, the connection of the three capacitors is delta. Moreover, we have:

$$V_L = 400 \, V, f = 50 \, Hz, Q = 600 \, kVAr \tag{1}$$

As we know, the reactive power of a single-phase capacitor can be determined as follows:

$$Q_{C,ph} = \frac{V_{rms,ph}^2}{X_c} = \omega C V_{rms,ph}^2 = 2\pi f C V_{rms,ph}^2 \tag{2}$$

Therefore, the reactive power of three-phase capacitor with the delta connection is:

$$Q_{C,3ph} = 3Q_{C,ph} = \frac{3V_{rms,ph}^2}{X_c} = 3\omega C V_{rms,ph}^2 = 6\pi f C V_{rms,ph}^2 \tag{3}$$

$$\Rightarrow C = \frac{Q_{C,3ph}}{6\pi f V_{rms,ph}^2} = \frac{600 \, kVAr}{6 \times 3.14 \times 50 \times 400^2} \Rightarrow C = 4000 \, \mu F$$

Choice (2) is the answer.

2.29. Based on the information given in the problem, we have the following specifications:

$$G : 22 \, kV, 90 \, MVA, X_G = 18\% \tag{1}$$

$$T_1 : 22/220 \, kV, 50 \, MVA, X_{T1} = 10\% \tag{2}$$

$$T_2 : 220/11 \, kV, 40 \, MW, X_{T2} = 6\% \tag{3}$$

$$T_3 : 22/110 \, kV, 40 \, MW, X_{T3} = 6.4\% \tag{4}$$

$$T_4 : 110/11 \, kV, 40 \, MW, X_{T4} = 8\% \tag{5}$$

$$M : 10.45 \, kV, 66.5 \, MVA, X_M = 18.5\% \tag{6}$$

$$TL_1 : 220 \, kV, 48.4 \, \Omega \tag{7}$$

$$TL_2 : 110 \, kV, 65.5 \, \Omega \tag{8}$$

$$S_B = 100 \, MVA, V_{B1} = 22 \, kV \tag{9}$$

As we know, base MVA is applied for the whole power system; however, base voltage might be different in each zone. Figure 2.19.2 shows the zones with the related base voltages that can be determined as follows:

$$V_{B2} = 22 \, kV \times \frac{220}{22} = 220 \, kV \tag{10}$$

$$V_{B3} = 220 \, kV \times \frac{11}{220} = 11 \, kV \tag{11}$$

$$V_{B4} = 22 \, kV \times \frac{110}{22} = 110 \, kV \tag{12}$$

Now, we need to update the per unit (p.u.) value of the impedances based on the new base MVA and voltages as follows:

$$X_{new,p.u.} = X_{old,p.u.} \times \frac{S_{B,new}}{S_{B,old}} \times \left(\frac{V_{B,old}}{V_{B,new}}\right)^2 \tag{13}$$

$$X_{G,new,p.u.} = 0.18 \times \frac{100}{90} \times \left(\frac{22}{22}\right)^2 = 0.2 \, p.u. \tag{14}$$

$$X_{T1,new,p.u.} = 0.10 \times \frac{100}{50} \times \left(\frac{22}{22}\right)^2 = 0.2 \, p.u. \tag{15}$$

$$X_{T2,new,p.u.} = 0.06 \times \frac{100}{40} \times \left(\frac{220}{220}\right)^2 = 0.15 \ p.u. \tag{16}$$

$$X_{T3,new,p.u.} = 0.064 \times \frac{100}{40} \times \left(\frac{11}{11}\right)^2 = 0.16 \ p.u. \tag{17}$$

$$X_{T4,new,p.u.} = 0.08 \times \frac{100}{40} \times \left(\frac{110}{110}\right)^2 = 0.2 \ p.u. \tag{18}$$

$$X_{M,new,p.u.} = 0.185 \times \frac{100}{66.5} \times \left(\frac{10.45}{11}\right)^2 = 0.25 \ p.u. \tag{19}$$

The impedance of the lines has been given in Ohms. Therefore, to present them in per unit, we need to determine the base impedance for their zones as follows:

$$S_B = \frac{(V_B)^2}{Z_B} \Rightarrow Z_B = \frac{(V_B)^2}{S_B} \tag{20}$$

$$Z_{B2} = \frac{(220 \ kV)^2}{100 \ MVA} = 484 \ \Omega \tag{21}$$

$$Z_{B4} = \frac{(110 \ kV)^2}{100 \ MVA} = 121 \ \Omega \tag{22}$$

Therefore:

$$X_{TL1,new,p.u.} = \frac{48.4}{484} = 0.1 \ p.u. \tag{23}$$

$$X_{TL2,new,p.u.} = \frac{65.5}{121} = 0.5 \ p.u. \tag{24}$$

Now, by using (14)–(19) and ((23)) –((24)), we can draw the impedance diagram of the power system which is illustrated in Fig. 2.19.3. Since the equivalent impedance is requested, the power supply of the generator and the EMF of the motor are turned off (short-circuited) in the diagram.

The impedance seen from the first bus can be calculated as follows:

$$\mathbf{Z}_{eq,p.u.} = (\ j0.2) \Big\| \Big((\ j0.2 + j0.1 + j0.15) \big\| (\ j0.16 + j0.5 + j0.2) + j0.25 \Big)$$

$$= (\ j0.2) \Big\| \Big((\ j0.45) \big\| (\ j0.86) + j0.25 \Big) = (\ j0.2) \Big\| \left(\frac{(\ j0.45)(\ j0.86)}{(\ j0.45) + (\ j0.86)} + j0.25 \right)$$

$$\approx (\ j0.2) \Big\| (\ j0.55) = \frac{(\ j0.2)(\ j0.55)}{(\ j0.2) + (\ j0.55)}$$

$$\mathbf{Z}_{eq,p.u.} \approx j0.14 \ p.u.$$

Choice (1) is the answer.

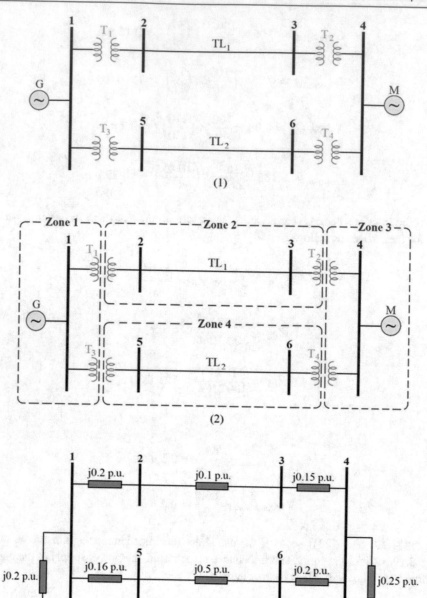

Fig. 2.19 The power system of solution of problem 2.29

2.30. Based on the information given in the problem, we have the following specifications:

$$G : 100 \ V \tag{1}$$

$$T_1 : 200/400 \ V, 1 \ kVA, X_{T1} = 0.1 \ p.u. \tag{2}$$

$$\text{Line} : \mathbf{Z}_{\text{Line}} = j8 \ \Omega \tag{3}$$

$$T_2 : 200/200 \ V, 2 \ kVA, X_{T2} = 0.1 \ p.u. \tag{4}$$

$$\text{Load}: \mathbf{Z_{Load}} = j6 \ \Omega \tag{5}$$

$$V_{B1} = 100 \ V, S_B = 1 \ kVA \tag{6}$$

The base voltage in the second and third zones can be calculated as follows:

$$V_{B2} = 100 \ V \times \frac{400}{200} = 200 \ V \tag{7}$$

$$V_{B3} = 100 \ V \times \frac{400}{200} \times \frac{200}{200} = 200 \ V \tag{8}$$

Now, we need to update the per unit (p.u.) value of the impedances based on the new base MVA and voltages as follows:

$$X_{new,p.u.} = X_{old,p.u.} \times \frac{S_{B,new}}{S_{B,old}} \times \left(\frac{V_{B,old}}{V_{B,new}}\right)^2 \tag{9}$$

$$X_{T1,new,p.u.} = 0.1 \times \frac{1}{1} \times \left(\frac{200}{100}\right)^2 = 0.4 \ p.u. \tag{10}$$

$$X_{T2,new,p.u.} = 0.1 \times \frac{1}{2} \times \left(\frac{200}{200}\right)^2 = 0.05 \ p.u. \tag{11}$$

To present the impedance of the line in per unit (p.u.) value, we need to determine the base impedance in the second zone, as follows:

$$S_B = \frac{(V_{B2})^2}{Z_{B2}} \Rightarrow Z_{B2} = \frac{(V_{B2})^2}{S_B} = \frac{(200 \ V)^2}{1 \ kVA} = 40 \ \Omega \tag{12}$$

$$\mathbf{Z_{Line,p.u.}} = \frac{\mathbf{Z_{Line}}}{Z_{B2}} \Rightarrow \mathbf{Z_{Line,p.u.}} = \frac{j8}{40} = j0.2 \ p.u. \tag{13}$$

Likewise, to present the impedance of the load in per unit value, we need to determine the base impedance in the third zone, as follows:

$$S_B = \frac{(V_{LB})^2}{Z_B} \Rightarrow Z_{B3} = \frac{(V_{LB3})^2}{S_B} = \frac{(200 \ V)^2}{1 \ kVA} = 40 \ \Omega \tag{14}$$

$$\mathbf{Z_{Load,p.u.}} = \frac{\mathbf{Z_{Load}}}{Z_{B3}} \Rightarrow \mathbf{Z_{Load,p.u.}} = \frac{j6}{40} = j0.15 \ p.u. \tag{15}$$

The voltage of the generator in per unit can be determined, as follows:

$$\mathbf{V_{G,p.u.}} = \frac{\mathbf{V_G}}{V_{B1}} \Rightarrow \mathbf{V_{G,p.u.}} = \frac{100}{100} = 1 \ p.u. \tag{16}$$

Figure 2.20.3 shows the impedance diagram of the power system based on (10), (11), (13), (15), and (16). The current can be calculated as follows:

$$\mathbf{I_{p.u.}} = \frac{\mathbf{V_{G,p.u.}}}{\mathbf{Z_{Total,p.u.}}} = \frac{1}{j0.4 + j0.2 + j0.05 + j0.15} \Rightarrow \mathbf{I_{p.u.}} = -j1.25 \ p.u. \Rightarrow |I_{p.u.}| = 1.25 \ p.u.$$

Choice (4) is the answer.

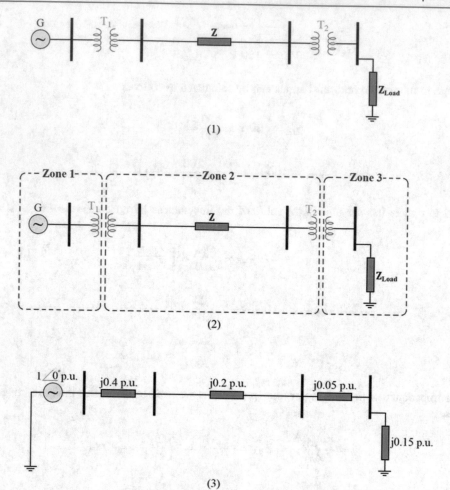

Fig. 2.20 The power system of solution of problem 2.30

Problems: Transmission Line Parameters 3

Abstract

In this chapter, the problems concerning with transmission line parameters are presented. The subjects include the Geometrical Mean Distance (GMD) and Geometrical Mean Radius (GMR) of conductors and the inductance and capacitance of single-phase and three-phase transmission lines bundled with a variety of arrangements. In this chapter, the problems are categorized in different levels based on their difficulty levels (easy, normal, and hard) and calculation amounts (small, normal, and large). Additionally, the problems are ordered from the easiest problem with the smallest computations to the most difficult problems with the largest calculations.

3.1. What is the main purpose of conductors bundling in transmission lines?

Difficulty level ● Easy ○ Normal ○ Hard
Calculation amount ● Small ○ Normal ○ Large

1) Decreasing inductive reactance of transmission line
2) Decreasing resistance of transmission line
3) Decreasing Corona power loss by reducing effective electric filed around conductors
4) Decreasing Corona power loss by reducing effective magnetic field around conductors

3.2. Determine the Geometrical Mean Radius (GMR) of the conductors with the arrangements shown in Fig. 3.1. The Geometrical Mean Radius (GMR) of each conductor is r'.

Difficulty level ● Easy ○ Normal ○ Hard
Calculation amount ● Small ○ Normal ○ Large

1) $r'\sqrt{D}$
2) $D\sqrt{r'}$
3) $\sqrt{r'D}$
4) $\sqrt{\frac{r'}{D}}$

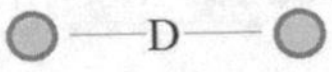

Fig. 3.1 The power system of problem 3.2

3.3. Which one of the following choices is correct about the effect of bundling of conductors of a transmission line on its inductance, capacitance, and characteristic impedance?

Difficulty level ● Easy ○ Normal ○ Hard
Calculation amount ● Small ○ Normal ○ Large

M. Rahmani-Andebili, *Power System Analysis*, https://doi.org/10.1007/978-3-030-84767-8_3

1) Decrease, decrease, no change
2) Increase, decrease, increase
3) Decrease, increase, decrease
4) Increase, increase, no change

3.4. Determine the Geometrical Mean Radius (GMR) of the conductors with the arrangements shown in Fig. 3.2. The Geometrical Mean Radius (GMR) of each conductor is r'.

Difficulty level   ○ Easy   ● Normal   ○ Hard
Calculation amount   ● Small   ○ Normal   ○ Large

1) $\sqrt[8]{2r'^6 D^6}$
2) $\sqrt[4]{2^3 r'^3 D}$
3) $\sqrt[4]{2^2 r' D}$
4) $\sqrt[8]{2r'^2 D^6}$

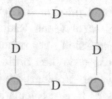

Fig. 3.2 The power system of problem 3.4

3.5. Determine the Geometrical Mean Radius (GMR) of the conductors with the arrangements shown in Fig. 3.3. The radius of each conductor is r.

Difficulty level   ○ Easy   ● Normal   ○ Hard
Calculation amount   ● Small   ○ Normal   ○ Large

1) $1.722r$
2) $1.834r$
3) $1.725r$
4) $1.532r$

Fig. 3.3 The power system of problem 3.5

3.6. Figure 3.4 shows a single-phase transmission line including two conductors ("1" and "3") for sending power and one conductor ("2") for receiving power. The Geometrical Mean Radius (GMR) of each conductor is r'. Calculate the inductance of the line in H/m.

Difficulty level   ○ Easy   ● Normal   ○ Hard
Calculation amount   ○ Small   ● Normal   ○ Large

1) $2 \times 10^{-7} \ln\left(\frac{D}{r'}\right)$
2) $2 \times 10^{-7} \ln\left(\sqrt{\frac{r'}{2D}}\right)$
3) $2 \times 10^{-7} \ln\left(\sqrt{\frac{D}{2r'}}\right)$
4) $10^{-7}\left(3\ln\left(\frac{D}{r'}\right) - \ln(2)\right)$

Fig. 3.4 The power system of problem 3.6

3.7. Figure 3.5 shows a single-phase transmission line. Herein, conductor "1" is for sending power, and conductors "2" and "3" are for receiving power. The Geometrical Mean Radius (GMR) of each conductor is r'. Calculate the inductance of the line in H/m.

Difficulty level   ○ Easy   ● Normal   ○ Hard
Calculation amount   ○ Small   ● Normal   ○ Large

1) $6 \times 10^{-7} \ln \left(\frac{D}{r'} \right)$

2) $2 \times 10^{-7} \ln \left(\frac{D}{r'} \right)$

3) $3 \times 10^{-7} \ln \left(\frac{D}{r'} \right)$

4) $4 \times 10^{-7} \ln \left(\frac{D}{r'} \right)$

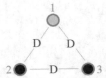

Fig. 3.5 The power system of problem 3.7

3.8. What difference can we see in the capacitance of a transmission line if we change the conductor arrangements from the two-bundling to the three-bundling, as can be seen in Fig. 3.6? The Geometrical Mean Radius (GMR) of each conductor is r' and $D > r'$. Herein, the distance between the phases is not changed.

Difficulty level   ○ Easy   ● Normal   ○ Hard
Calculation amount   ○ Small   ● Normal   ○ Large

1) It will decrease.

2) It will not change.

3) It will increase.

4) It can decrease or increase.

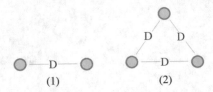

Fig. 3.6 The power system of problem 3.8

3.9. Figure 3.7 illustrates two single-phase transmission lines. The Geometrical Mean Radius (GMR) of each conductor is r'. In Fig. 3.7 (b), conductors "2" and "3" are for sending power, and conductor "1" is for receiving power. What relation should be held between D and r' so that the inductances of the transmission lines become equal?

Difficulty level   ○ Easy   ● Normal   ○ Hard
Calculation amount   ○ Small   ● Normal   ○ Large

1) $D = \frac{9}{4} r'$

2) $D = \frac{37}{16} r'$

3) $D = \frac{5}{2} r'$

4) $D = \frac{21}{8} r'$

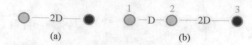

Fig. 3.7 The power system of problem 3.9

3.10. Figure 3.8 shows a single-phase line including two conductors ("2" and "3") for sending and one conductor ("1") for receiving power. The Geometrical Mean Radius (GMR) of each conductor is r'. Calculate the capacitance of the line in F/m.

Difficulty level   ○ Easy   ○ Normal   ● Hard
Calculation amount   ○ Small   ● Normal   ○ Large

1) $\frac{4\pi\varepsilon_0}{2\ln\left(\frac{D}{r'}\right)}$

2) $\frac{2\pi\varepsilon_0}{3\ln\left(\frac{2D}{r'}\right)}$

3) $\frac{4\pi\varepsilon_0}{3\ln\left(\frac{D}{2r'}\right)}$

4) $\frac{4\pi\varepsilon_0}{3\ln\left(\frac{D}{r'}\right)}$

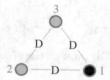

Fig. 3.8 The power system of problem 3.10

3.11. Figure 3.9 illustrates two three-phase transmission lines. The Geometrical Mean Radius (GMR) of each conductor is r' and $r' < d$. What relation should be held between d and r' so that the inductance of the transmission lines become equal?

Difficulty level   ○ Easy   ○ Normal   ● Hard
Calculation amount   ○ Small   ● Normal   ○ Large

1) $d = \frac{r'}{\sqrt{2}}$.

2) $d = 2r'$.

3) $d = \sqrt{2}r'$.

4) No possible relation can be found.

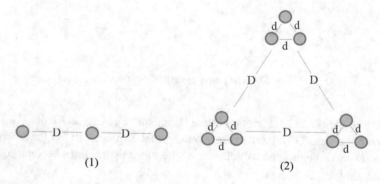

Fig. 3.9 The power system of problem 3.11

3.12. Which one of the arrangements of a three-phase transmission line, shown in Fig. 3.10, has the least inductance and the most capacitance? The Geometrical Mean Radius (GMR) of each conductor is r'.

Difficulty level   ○ Easy   ○ Normal   ● Hard
Calculation amount   ○ Small   ○ Normal   ● Large

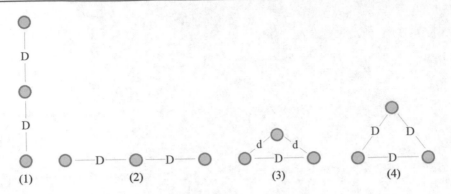

Fig. 3.10 The power system of problem 3.12

3.13. What difference can we see in the inductance of a transmission line if we change the conductor arrangements from the two-bundling to three-bundling, as can be seen in Fig. 3.11? The Geometrical Mean Radius (GMR) of each conductor is r' and $D = 4r'$. Herein, the distance between the phases is kept constant.

Difficulty level   ○ Easy   ○ Normal   ● Hard
Calculation amount   ○ Small   ○ Normal   ● Large

1) A decrease about $\frac{2}{3} \times 10^{-7} \ln(2)$

2) No change

3) An increase about $\frac{2}{3} \times 10^{-7} \ln(2)$

4) An increase about $\frac{3}{2} \times 10^{-7} \ln(2)$

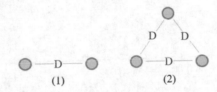

Fig. 3.11 The power system of problem 3.13

4 Solutions of Problems: Transmission Line Parameters

Abstract
In this chapter, the problems of the third chapter are fully solved, in detail, step by step, and with different methods.

4.1. Decreasing Corona power loss is the main purpose of conductors bundling in transmission lines which is caused by reducing effective electric filed around conductors. Choice (3) is the answer.

4.2. Based on the information given in the problem, we know that the Geometrical Mean Radius (GMR) of each conductor is r'.

Therefore:

$$GMR = \sqrt[2\times 2]{(r' \times D) \times (r' \times D)} \Rightarrow GMR = \sqrt{r'D}$$

Choice (3) is the answer.

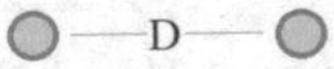

Fig. 4.1 The power system of solution of problem 4.2

4.3. As we know, the inductance and capacitance of a transmission line can be determined as follows:

$$L \downarrow = 2 \times 10^{-7} \ln \left(\frac{GMD}{GMR \uparrow} \right)$$

$$C \uparrow = \frac{2\pi\varepsilon_0}{\ln \left(\frac{GMD}{GMR \uparrow} \right)}$$

Bundling of conductors of a transmission line can increase its Geometrical Mean Radius (GMR). Therefore, the inductance and the capacitance of the transmission line will decrease and increase, respectively. Moreover, based on the relation below, the characteristic impedance will decrease:

$$Z_C \downarrow = \sqrt{\frac{L \downarrow}{C \uparrow}}$$

Choice (3) is the answer.

M. Rahmani-Andebili, *Power System Analysis*, https://doi.org/10.1007/978-3-030-84767-8_4

4.4. Based on the information given in the problem, we know that the Geometrical Mean Radius (GMR) of each conductor is r'. The Geometrical Mean Radius (GMR) of the bundled conductors can be determined as follows:

$$GMR = \sqrt[4\times4]{(D_{11}D_{12}D_{13}D_{14}) \times (D_{22}D_{21}D_{23}D_{24}) \times (D_{33}D_{31}D_{32}D_{34}) \times (D_{44}D_{41}D_{42}D_{43})}$$

$$\sqrt[4\times4]{(D_{11}D_{12}D_{13}D_{14})^4} = \sqrt[4]{D_{11}D_{12}D_{13}D_{14}} = \sqrt[4]{r' \times D \times D \times D\sqrt{2}} = \sqrt[4]{r'D^3\sqrt{2}}$$

$$GMR = \sqrt[8]{2r'^2 D^6}$$

Choice (4) is the answer.

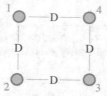

Fig. 4.2 The power system of solution of problem 4.4

4.5. Based on the information given in the problem, we know that the radius of each conductor is r. Therefore, the Geometrical Mean Radius (GMR) of each conductor is:

$$r' = re^{-\frac{1}{4}}$$

Therefore, the Geometrical Mean Radius (GMR) of the bundled conductors is:

$$GMR = \sqrt[4\times4]{(D_{11}D_{12}D_{13}D_{14}) \times (D_{22}D_{21}D_{23}D_{24}) \times (D_{33}D_{31}D_{32}D_{34}) \times (D_{44}D_{41}D_{42}D_{43})}$$

$$\sqrt[4\times4]{(D_{11}D_{12}D_{13}D_{14})^4} = \sqrt[4]{D_{11}D_{12}D_{13}D_{14}} = \sqrt[4]{r' \times 2r \times 2r \times 2r\sqrt{2}} = \sqrt[4]{e^{-\frac{1}{4}}2^{\frac{7}{2}}r^4} = e^{-\frac{1}{16}}2^{\frac{7}{8}}r$$

$$GMR = 1.722r$$

Choice (2) is the answer.

Fig. 4.3 The power system of solution of problem 4.5

4.6. Based on the information given in the problem, we know that conductors "1" and "3" are for sending power and conductor "2" is for receiving power. Moreover, the radius of each conductor is r'.

To calculate the inductance of a single-phase transmission line, we need to calculate the sum of the inductances of power sending and power receiving lines, as they are connected in series. Therefore:

$$L_{Total} = L_{13} + L_2 = 2 \times 10^{-7} \ln\left(\frac{GMD_{13}}{GMR_{13}}\right) + 2 \times 10^{-7} \ln\left(\frac{GMD_2}{GMR_2}\right) \tag{1}$$

$$GMD_{13} = GMD_2 = \sqrt{D \times D} = D \tag{2}$$

$$GMR_{13} = \sqrt[2 \times 2]{r' \times 2D \times r' \times 2D} = \sqrt{2Dr'} \tag{3}$$

$$GMR_2 = r' \tag{4}$$

Solving (1)–(4):

$$L_{Total} = 2 \times 10^{-7} \ln\left(\frac{D}{\sqrt{2Dr'}}\right) + 2 \times 10^{-7} \ln\left(\frac{D}{r'}\right) = 2 \times 10^{-7} \ln\left(\frac{D}{\sqrt{2Dr'}} \times \frac{D}{r'}\right)$$

$$= 2 \times 10^{-7} \ln\left(\frac{D^{\frac{3}{2}}}{2^{\frac{1}{2}}r'^{\frac{3}{2}}}\right) = 2 \times 10^{-7} \left(\ln\left(\frac{D}{r'}\right)^{\frac{3}{2}} + \ln\left(\frac{1}{2^{\frac{1}{2}}}\right)\right)$$

$$L_{Total} = 10^{-7}\left(3\ln\left(\frac{D}{r'}\right) - \ln(2)\right)$$

Choice (4) is the answer.

Fig. 4.4 The power system of solution of problem 4.6

4.7. Based on the information given in the problem, we know that conductor "1" is for sending power and conductors "2" and "3" are for receiving power. Moreover, the Geometrical Mean Radius (GMR) of each conductor is r'.

To calculate the inductance of a single-phase transmission line, we need to individually calculate the inductances of power sending line and power receiving line and then add them up, as they are in series. Therefore:

$$L_{Total} = L_1 + L_{23} = 2 \times 10^{-7} \ln\left(\frac{GMD_1}{GMR_1}\right) + 2 \times 10^{-7} \ln\left(\frac{GMD_{23}}{GMR_{23}}\right) \tag{1}$$

$$GMD_1 = GMD_{23} = \sqrt{D \times D} = D \tag{2}$$

$$GMR_1 = r' \tag{3}$$

$$GMR_{23} = \sqrt[4]{r' \times D \times r' \times D} = \sqrt{r'D} \tag{4}$$

Solving (1)–(4):

$$L_{Total} = 2 \times 10^{-7} \ln\left(\frac{D}{r'}\right) + 2 \times 10^{-7} \ln\left(\frac{D}{\sqrt{r'D}}\right) = 2 \times 10^{-7} \ln\left(\frac{D}{\sqrt{r'D}} \times \frac{D}{r'}\right) = 2 \times 10^{-7} \ln\left(\frac{D}{r'}\right)^{\frac{3}{2}}$$

$$L_{Total} = 3 \times 10^{-7} \ln\left(\frac{D}{r'}\right)$$

Choice (3) is the answer.

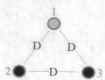

Fig. 4.5 The power system of solution of problem 4.7

4.8. Based on the information given in the problem, we know that the Geometrical Mean Radius (GMR) of each conductor is r'.

As we know, the capacitance of a transmission line can be determined as follows:

$$C = \frac{2\pi\varepsilon_0}{\ln\left(\frac{GMD}{GMR}\right)}$$ (1)

Therefore:

$$C_{2b} = \frac{2\pi\varepsilon_0}{\ln\left(\frac{GMD_{2b}}{GMR_{2b}}\right)}$$ (2)

$$C_{3b} = \frac{2\pi\varepsilon_0}{\ln\left(\frac{GMD_{3b}}{GMR_{3b}}\right)}$$ (3)

Where:

$$GMR_{2b} = \sqrt[4]{(r' \times D)^2} = \sqrt{r'D}$$ (4)

$$GMR_{3b} = \sqrt[9]{(r' \times D \times D)^3} = \sqrt[3]{r'D^2}$$ (5)

The Geometrical Mean Distance (GMD) will not change, since only the bundling is changed. Therefore:

$$GMD_{2b} = GMD_{3b} = GMD$$ (6)

As can be noticed from (4) and (5):

$$GMR_{2b} < GMR_{3b} \xrightarrow{Using\ (6)} \ln\left(\frac{GMD}{GMR_{2b}}\right) > \ln\left(\frac{GMD}{GMR_{3b}}\right) \Rightarrow \frac{2\pi\varepsilon_0}{\ln\left(\frac{GMD_{2b}}{GMR_{2b}}\right)} < \frac{2\pi\varepsilon_0}{\ln\left(\frac{GMD_{3b}}{GMR_{3b}}\right)}$$

Therefore:

$$C_{2b} < C_{3b}$$

Choice (3) is the answer.

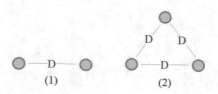

Fig. 4.6 The power system of solution of problem 4.8

4.9. Based on the information given in the problem, we know that the Geometrical Mean Radius (GMR) of each conductor is r'. Moreover, in Fig. 4.7 (b), conductors "2" and "3" are for sending power, and conductor "1" is for receiving power. In addition:

$$L_a = L_b \tag{1}$$

To calculate the inductance of a single-phase transmission line, we need to calculate the sum of the inductances of power sending line and power receiving line, as they are connected in series. Therefore:

$$L_a = 2 \times \left(2 \times 10^{-7} \ln \left(\frac{2D}{r'} \right) \right) = 2 \times 10^{-7} \ln \left(\frac{2D}{r'} \right)^2 \tag{2}$$

$$L_b = L_{12} + L_3 = 2 \times 10^{-7} \ln \left(\frac{\sqrt{3D \times 2D}}{\sqrt{r'D}} \right) + 2 \times 10^{-7} \ln \left(\frac{\sqrt{3D \times 2D}}{r'} \right) = 2 \times 10^{-7} \ln \left(\frac{6D^{\frac{3}{2}}}{r'^{\frac{3}{2}}} \right) \tag{3}$$

Solving (1)–(3):

$$2 \times 10^{-7} \ln \left(\frac{2D}{r'} \right)^2 = 2 \times 10^{-7} \ln \left(\frac{6D^{\frac{3}{2}}}{r'^{\frac{3}{2}}} \right) \Rightarrow \left(\frac{2D}{r'} \right)^2 = \frac{6D^{\frac{3}{2}}}{r'^{\frac{3}{2}}} \Rightarrow \frac{2D^{\frac{1}{2}}}{r'^{\frac{1}{2}}} = \frac{3}{1} \Rightarrow D = \frac{9}{4} r'$$

Choice (1) is the answer.

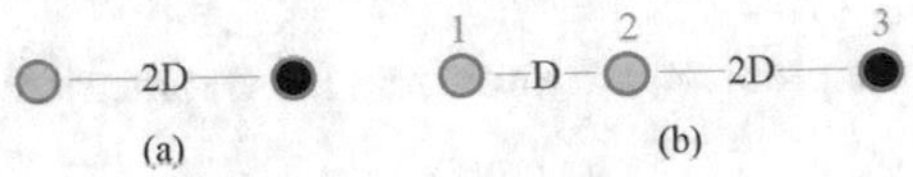

Fig. 4.7 The power system of solution of problem 4.9

4.10. Based on the information given in the problem, we know that conductors "2" and "3" are for sending power and conductor "1" is for receiving power. Moreover, the Geometrical Mean Radius (GMR) of each conductor is r'.

To calculate the capacitance of a single-phase transmission line, we need to determine the equivalent capacitance of the capacitance of power sending line and the capacitance of the power receiving line, since they are connected in series. Thus:

$$C_{Total} = \frac{C_1 C_{23}}{C_1 + C_{23}} \tag{1}$$

$$C_1 = \frac{2\pi\varepsilon_0}{\ln \left(\frac{GMD_1}{GMR_1} \right)} \tag{2}$$

$$C_{23} = \frac{2\pi\varepsilon_0}{\ln \left(\frac{GMD_{23}}{GMR_{23}} \right)} \tag{3}$$

Where:

$$GMD_1 = GMD_{23} = \sqrt{D \times D} = D \tag{4}$$

$$GMR_1 = r' \tag{5}$$

$$GMR_{23} = \sqrt[4]{r' \times D \times r' \times D} = \sqrt{r'D} \tag{6}$$

Solving (1)–(6):

$$C_{Total} = \frac{\frac{2\pi\varepsilon_0}{\ln\left(\frac{D}{r'}\right)} \frac{2\pi\varepsilon_0}{\ln\left(\frac{D}{\sqrt{r'D}}\right)}}{\frac{2\pi\varepsilon_0}{\ln\left(\frac{D}{r'}\right)} + \frac{2\pi\varepsilon_0}{\ln\left(\frac{D}{\sqrt{r'D}}\right)}} = \frac{\frac{\ln\left(\frac{D}{r'}\right)\ln\left(\frac{D}{\sqrt{r'D}}\right)}{\ln\left(\frac{D}{r'}\right) + \ln\left(\frac{D}{\sqrt{r'D}}\right)}}{\ln\left(\frac{D}{r'}\right)\ln\left(\frac{D}{\sqrt{r'D}}\right)} = \frac{2\pi\varepsilon_0}{\ln\left(\frac{D}{r'}\right) + \ln\left(\frac{D}{\sqrt{r'D}}\right)} = \frac{2\pi\varepsilon_0}{\ln\left(\frac{D}{r'}\right)^{\frac{3}{2}}}$$

$$C_{Total} = \frac{4\pi\varepsilon_0}{3\ln\left(\frac{D}{r'}\right)}$$

Choice (4) is the answer.

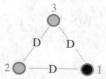

Fig. 4.8 The power system of solution of problem 4.10

4.11. Based on the information given in the problem, we know that the Geometrical Mean Radius (GMR) of each conductor is r' and:

$$r' < d \tag{1}$$

$$L_1 = L_2 \tag{2}$$

As we know, the inductance of a three-phase transmission line can be determined as follows:

$$L = 2 \times 10^{-7} \ln\left(\frac{GMD}{GMR}\right) \tag{3}$$

Therefore:

$$L_1 = 2 \times 10^{-7} \ln\left(\frac{GMD_1}{GMR_1}\right) \tag{4}$$

$$L_2 = 2 \times 10^{-7} \ln\left(\frac{GMD_2}{GMR_2}\right) \tag{5}$$

Where:

$$GMD_1 = \sqrt[3]{D \times D \times 2D} = D\sqrt[3]{2} \tag{6}$$

$$GMD_2 = \sqrt[3]{D \times D \times D} = D \tag{7}$$

$$GMR_1 = r' \tag{8}$$

$$GMR_2 = \sqrt[3]{r' \times d \times d} = \sqrt[3]{r'd^2} \tag{9}$$

Solving (2)–(9):

$$2 \times 10^{-7} \ln\left(\frac{D\sqrt[3]{2}}{r'}\right) = 2 \times 10^{-7} \ln\left(\frac{D}{\sqrt[3]{r'd^2}}\right) \Rightarrow \frac{D\sqrt[3]{2}}{r'} = \frac{D}{\sqrt[3]{r'd^2}} \Rightarrow r'^{\frac{2}{3}} = d^{\frac{2}{3}}2^{\frac{1}{3}} \Rightarrow r' = d\sqrt{2} \tag{10}$$

Equations (1) and (10) do not have any solution. Choice (4) is the answer.

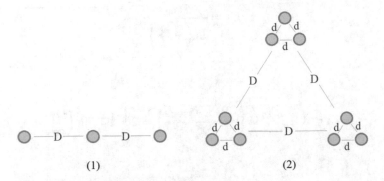

Fig. 4.9 The power system of solution of problem 4.11

4.12. As we know, the inductance and the capacitance of a three-phase transmission line can be determined as follows:

$$L = 2 \times 10^{-7} \ln\left(\frac{GMD}{GMR}\right) \tag{1}$$

$$C = \frac{2\pi\varepsilon_0}{\ln\left(\frac{GMD}{GMR}\right)} \tag{2}$$

Based on the information given in the problem, we know that the Geometrical Mean Radius (GMR) of each conductor is r'. The transmission lines are not bundled. Hence:

$$GMR_1 = GMR_2 = GMR_3 = GMR_4 = r' \tag{3}$$

Moreover, as can be noticed from Fig. 4.10.3, we have:

$$d < D \tag{4}$$

For case 1:

$$L_1 = 2 \times 10^{-7} \ln\left(\frac{\sqrt[3]{D \times D \times 2D}}{r'}\right) = 2 \times 10^{-7} \ln\left(\frac{D\sqrt[3]{2}}{r'}\right) \tag{5}$$

$$C_1 = \frac{2\pi\varepsilon_0}{\ln\left(\frac{D\sqrt[3]{2}}{r'}\right)} \tag{6}$$

For case 2:

$$L_2 = 2 \times 10^{-7} \ln\left(\frac{\sqrt[3]{D \times D \times 2D}}{r'}\right) = 2 \times 10^{-7} \ln\left(\frac{D\sqrt[3]{2}}{r'}\right) \tag{7}$$

$$C_2 = \frac{2\pi\varepsilon_0}{\ln\left(\frac{D\sqrt[3]{2}}{r'}\right)} \tag{8}$$

For case 3:

$$L_3 = 2 \times 10^{-7} \ln\left(\frac{\sqrt[3]{D \times d \times d}}{r'}\right) = 2 \times 10^{-7} \ln\left(\frac{\sqrt[3]{Dd^2}}{r'}\right) \tag{9}$$

$$C_3 = \frac{2\pi\varepsilon_0}{\ln\left(\frac{\sqrt[3]{Dd^2}}{r'}\right)} \tag{10}$$

For case 4:

$$L_4 = 2 \times 10^{-7} \ln\left(\frac{\sqrt[3]{D \times D \times D}}{r'}\right) = 2 \times 10^{-7} \ln\left(\frac{D}{r'}\right) \tag{11}$$

$$C_4 = \frac{2\pi\varepsilon_0}{\ln\left(\frac{D}{r'}\right)} \tag{12}$$

From (5) to (12), we can conclude that:

$$2 \times 10^{-7} \ln\left(\frac{\sqrt[3]{Dd^2}}{r'}\right) < 2 \times 10^{-7} \ln\left(\frac{D}{r'}\right) < 2 \times 10^{-7} \ln\left(\frac{D\sqrt[3]{2}}{r'}\right) \Rightarrow L_3 < L_4 < L_2 = L_1 \tag{13}$$

$$\frac{2\pi\varepsilon_0}{\ln\left(\frac{D\sqrt[3]{2}}{r'}\right)} < \frac{2\pi\varepsilon_0}{\ln\left(\frac{D}{r'}\right)} < \frac{2\pi\varepsilon_0}{\ln\left(\frac{\sqrt[3]{Dd^2}}{r'}\right)} \Rightarrow C_1 = C_2 < C_4 < C_3 \tag{14}$$

As can be seen in (13) and (14), arrangement 3 has the least inductance and the most capacitance. Choice (3) is the answer.

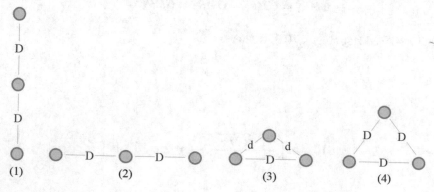

Fig. 4.10 The power system of solution of problem 4.12

4.13. Based on the information given in the problem, we know that the Geometrical Mean Radius (GMR) of each conductor is r' and:

$$D = 4r' \tag{1}$$

As we know, the inductance of a transmission line can be determined as follows:

$$L = 2 \times 10^{-7} \ln \left(\frac{GMD}{GMR} \right) \tag{2}$$

Therefore:

$$L_{2b} = 2 \times 10^{-7} \ln \left(\frac{GMD_{2b}}{GMR_{2b}} \right) \tag{3}$$

$$L_{3b} = 2 \times 10^{-7} \ln \left(\frac{GMD_{3b}}{GMR_{3b}} \right) \tag{4}$$

Where:

$$GMR_{2b} = \sqrt[4]{(r' \times D)^2} = \sqrt{r'D} \xrightarrow{Using \ (1)} GMR_{2b} = \sqrt{4r'^2} = 2r' \tag{5}$$

$$GMR_{3b} = \sqrt[9]{(r' \times D \times D)^3} = \sqrt[3]{r'D^2} \xrightarrow{Using \ (1)} GMR_{3b} = \sqrt[3]{16r'^3} = 2^{\frac{4}{3}}r' \tag{6}$$

The Geometrical Mean Distance (GMD) will not change, since only the bundling is changed. Therefore:

$$GMD_{2b} = GMD_{3b} \tag{7}$$

Therefore:

$$L_{3b} - L_{2b} = 2 \times 10^{-7} \ln \left(\frac{GMD_{3b}}{GMR_{3b}} \right) - 2 \times 10^{-7} \ln \left(\frac{GMD_{2b}}{GMR_{2b}} \right) = 2 \times 10^{-7} \ln \left(\frac{GMD_{3b}}{GMR_{3b}} \times \frac{GMR_{2b}}{GMD_{2b}} \right) \tag{8}$$

Solving (7)–(8):

$$L_{3b} - L_{2b} = 2 \times 10^{-7} \ln \left(\frac{GMR_{2b}}{GMR_{3b}} \right) \tag{9}$$

Solving (5), (6), and (9):

$$L_{3b} - L_{2b} = 2 \times 10^{-7} \left(\ln \left(\frac{2r'}{2^{\frac{4}{3}}r'} \right) \right) = 2 \times 10^{-7} \left(\ln \left(2^{-\frac{1}{3}} \right) \right)$$

$$L_{3b} - L_{2b} = -\frac{2}{3} \times 10^{-7} \ln (2)$$

Choice (1) is the answer.

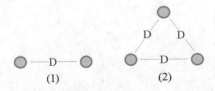

Fig. 4.11 The power system of solution of problem 4.13

Problems: Transmission Line Model and Performance

Abstract

In this chapter, the problems concerning with the transmission line model and performance are presented. The subjects include transmission line models, transmission line voltage regulation, transmission line compensation, and features of transmission matrix. In this chapter, the problems are categorized in different levels based on their difficulty levels (easy, normal, and hard) and calculation amounts (small, normal, and large). Additionally, the problems are ordered from the easiest problem with the smallest computations to the most difficult problems with the largest calculations.

5.1. Which one of the parameters below can be ignored for a short transmission line?

| Difficulty level | ● Easy | ○ Normal | ○ Hard |
| Calculation amount | ● Small | ○ Normal | ○ Large |

1) Resistance
2) Inductance
3) Reactance
4) Capacitance

5.2. Based on Ferranti effect, which one of the following terms is correct?

| Difficulty level | ● Easy | ○ Normal | ○ Hard |
| Calculation amount | ● Small | ○ Normal | ○ Large |

1) The voltage in the receiving end increases when the transmission line is operated in no-load or low-load conditions.
2) The voltage in the receiving end increases when the transmission line is operated in full-load condition.
3) The voltage in the receiving end increases when the transmission line is short-circuited.
4) The voltage in the receiving end decreases when the transmission line is operated in full-load condition.

5.3. Which one of the matrices below belongs to a transmission matrix of a real transmission line?

| Difficulty level | ● Easy | ○ Normal | ○ Hard |
| Calculation amount | ● Small | ○ Normal | ○ Large |

1) $\begin{bmatrix} j & 1 \\ 0 & -j \end{bmatrix}$

2) $\begin{bmatrix} 1 & j \\ 2 & 1 \end{bmatrix}$

3) $\begin{bmatrix} 1 & 2 \\ 3 & 1 \end{bmatrix}$

4) $\begin{bmatrix} 1 & j \\ 0 & 1 \end{bmatrix}$

5.4. Two power systems have the transmission matrices below. If these systems are cascaded, determine their equivalent transmission matrix:

$$[T_1] = \begin{bmatrix} 1 & j2 \\ 0 & 1 \end{bmatrix}, [T_2] = \begin{bmatrix} 1 & 0 \\ j2 & 1 \end{bmatrix}$$

Difficulty level   ● Easy   ○ Normal   ○ Hard
Calculation amount   ● Small   ○ Normal   ○ Large

1) $\begin{bmatrix} -3 & j2 \\ j2 & 1 \end{bmatrix}$

2) $\begin{bmatrix} 5 & j2 \\ j2 & 1 \end{bmatrix}$

3) $\begin{bmatrix} 2 & j2 \\ j2 & 2 \end{bmatrix}$

4) None of them

5.5. Calculate the characteristic impedance of a long lossless transmission line that has the inductance and capacitance of about 1 *mH/meter* and 10 *μF/meter*, respectively.

Difficulty level   ○ Easy   ● Normal   ○ Hard
Calculation amount   ● Small   ○ Normal   ○ Large

1) 20 Ω
2) 5 Ω
3) 10 Ω
4) 40 Ω

5.6. At the end of a transmission line with the characteristic impedance of $Z_C = (1 - j)\ \Omega$, a load with the impedance of $Z_L = (1 + j)\ \Omega$ has been connected. Which one of the following components needs to be installed in parallel to the load to remove the reflected waves of the voltage and current?

Difficulty level   ○ Easy   ● Normal   ○ Hard
Calculation amount   ● Small   ○ Normal   ○ Large

1) A capacitor with the reactance of 0.5
2) A capacitor with the reactance of 1
3) An inductor with the reactance of 0.5
4) An inductor with the reactance of 1

5.7. As is shown in Fig. 5.1, a medium transmission line has been presented by its T model. Calculate the charging current of the line ($I_{Charging}$).

Difficulty level   ○ Easy   ● Normal   ○ Hard
Calculation amount   ○ Small   ● Normal   ○ Large

1) Only $ZV_R(1 + 0.5YZ)^{-1}$
2) Only $YV_S(1 + 0.5YZ)^{-1}$
3) YV_R or $YV_S(1 + 0.5YZ)^{-1}$
4) Only $YV_R(1 + 0.5YZ)^{-1}$

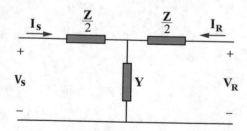

Fig. 5.1 The power system of problem 5.7

5.8. Figure 5.2 shows the single-line diagram of a short transmission line. Determine its transmission matrix.

Difficulty level   ○ Easy   ● Normal   ○ Hard
Calculation amount   ○ Small   ● Normal   ○ Large

1) $\begin{bmatrix} 1 + \mathbf{YZ} & 1 \\ \mathbf{Z} & \mathbf{Y} \end{bmatrix}$

2) $\begin{bmatrix} 1 + \mathbf{YZ} & \mathbf{Z} \\ \mathbf{Y} & 1 \end{bmatrix}$

3) $\begin{bmatrix} 1 + \mathbf{YZ} & 1 \\ \mathbf{Y} & \mathbf{Z} \end{bmatrix}$

4) $\begin{bmatrix} 1 + \mathbf{YZ} & \mathbf{Y} \\ \mathbf{Z} & 1 \end{bmatrix}$

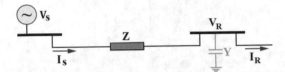

Fig. 5.2 The power system of problem 5.8

5.9. Determine the characteristic impedance of a transmission line that the relation below is true for its parameters:

$$\frac{R}{L} = \frac{G}{C}$$

Difficulty level   ○ Easy   ● Normal   ○ Hard
Calculation amount   ○ Small   ● Normal   ○ Large

1) $\frac{R}{L}$.

2) ∞.

3) 0.

4) It is equal to the characteristic impedance of a lossless transmission line.

5.10. Figure 5.3 shows the single-line diagram of a short transmission line that a resistor with the resistance of R has been installed in its middle point. Determine its transmission matrix.

Difficulty level   ○ Easy   ● Normal   ○ Hard
Calculation amount   ○ Small   ○ Normal   ● Large

1) $\begin{bmatrix} R + \mathbf{Z} & \mathbf{Z}(R + \mathbf{Z}) \\ \mathbf{Z} & R + \mathbf{Z} \end{bmatrix}$

2) $\begin{bmatrix} 1 + \mathbf{Z} & \mathbf{Z}(R + \mathbf{Z}) \\ \mathbf{Z} & R + \mathbf{Z} \end{bmatrix}$

3) $\begin{bmatrix} 1 + \dfrac{\mathbf{Z}}{2R} & \mathbf{Z}\left(R + \dfrac{\mathbf{Z}}{4R}\right) \\ \dfrac{1}{R} & 1 + \dfrac{\mathbf{Z}}{2R} \end{bmatrix}$

4) $\begin{bmatrix} 1 + \dfrac{\mathbf{Z}}{2R} & \mathbf{Z}\left(R + \dfrac{\mathbf{Z}}{2R}\right) \\ \dfrac{1}{R} & 1 + \dfrac{\mathbf{Z}}{4R} \end{bmatrix}$

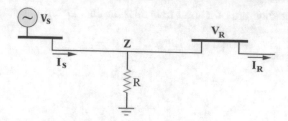

Fig. 5.3 The power system of problem 5.10

5.11. Calculate the characteristic impedance of a long transmission line that its transmission matrix is as follows:

$$[T] = \begin{bmatrix} \frac{1}{2} & j \\ \frac{3}{4}j & \frac{1}{2} \end{bmatrix}$$

| Difficulty level | ○ Easy | ○ Normal | ● Hard |
| Calculation amount | ○ Small | ● Normal | ○ Large |

1) $\frac{\sqrt{3}}{3}\ \Omega$

2) $\frac{2\sqrt{3}}{3}\ \Omega$

3) $\frac{3}{4}\ \Omega$

4) $\frac{1}{2}\ \Omega$

5.12. Calculate the charging current ($\mathbf{I_{Charging}}$) of a long transmission line.

| Difficulty level | ○ Easy | ○ Normal | ● Hard |
| Calculation amount | ○ Small | ● Normal | ○ Large |

1) $\frac{\mathbf{V_S}\tanh(\gamma l)}{\mathbf{Z_c}}$

2) $\mathbf{Z_c V_S}\tanh(\gamma l)$

3) $\frac{\mathbf{V_S}\coth(\gamma l)}{\mathbf{Z_c}}$

4) $\mathbf{Z_c V_S}\coth(\gamma l)$

5.13. In a long transmission line, consider the definitions below, and choose the correct relation between $\mathbf{Z_C}$, $\mathbf{Z_{S.C.}}$, and $\mathbf{Z_{O.C.}}$.
$\mathbf{Z_C}$: Characteristic impedance
$\mathbf{Z_{S.C.}}$: The impedance seen from the beginning of the transmission line if its end is short circuit
$\mathbf{Z_{O.C.}}$: The impedance seen from the beginning of the transmission line if its end is open circuit

| Difficulty level | ○ Easy | ○ Normal | ● Hard |
| Calculation amount | ○ Small | ● Normal | ○ Large |

1) $\mathbf{Z_C} = \mathbf{Z_{S.C.}} - \mathbf{Z_{O.C.}}$

2) $\mathbf{Z_C} = \mathbf{Z_{O.C.}} - \mathbf{Z_{S.C.}}$

3) $\mathbf{Z_C} = \sqrt{\mathbf{Z_{S.C.}Z_{O.C.}}}$

4) $\mathbf{Z_C} = \frac{1}{2}(\mathbf{Z_{S.C.}} + \mathbf{Z_{O.C.}})$

5.14. In a long transmission line, the impedance measured from the beginning of the line, when its end is open circuit, is the reciprocal of the impedance measured from the beginning of the line, when its end is short circuit. Which one of the following relations is correct among the parameters of the transmission matrix of this line?

$$[T] = \begin{bmatrix} A & B \\ C & D \end{bmatrix}$$

| Difficulty level | ○ Easy | ○ Normal | ● Hard |
| Calculation amount | ○ Small | ● Normal | ○ Large |

1) $A + B = \frac{1}{A-B}$

2) $A + B = \frac{1}{B-A}$

3) $A = \sqrt{1 - B^2}$

4) $A = \sqrt{B^2 - 1}$

5.15. In a no-load and lossless transmission line, which one of the following relations is correct? Herein, $\mathbf{V_R}$, $\mathbf{V_S}$, β, γ, and l are the voltage of receiving end, voltage of sending end, phase constant, propagation constant, and length of line, respectively.

Difficulty level   ○ Easy   ○ Normal   ● Hard
Calculation amount   ○ Small   ● Normal   ○ Large

1) $\mathbf{V_R} = \frac{\mathbf{V_S}}{\sin(\beta l)}$

2) $\mathbf{V_R} = \frac{\mathbf{V_S}}{\sin(\gamma l)}$

3) $\mathbf{V_R} = \frac{\mathbf{V_S}}{\cos(\beta l)}$

4) $\mathbf{V_R} = \frac{\mathbf{V_S}}{\cos(\gamma l)}$

5.16. A factory is supplied by an ideal transformer through a short transmission line. At the bus of the factory, a shunt capacitor has been installed to correct its power factor. Which one of the transmission matrices below is correct for this power system?

Difficulty level   ○ Easy   ○ Normal   ● Hard
Calculation amount   ○ Small   ○ Normal   ● Large

1) $\begin{bmatrix} \frac{1}{a} & 0 \\ 0 & a \end{bmatrix} \begin{bmatrix} 1 + \mathbf{ZY} & \mathbf{Z} \\ \mathbf{Y} & 1 \end{bmatrix}$

2) $\begin{bmatrix} a & 0 \\ 0 & \frac{1}{a} \end{bmatrix} \begin{bmatrix} 1 + \mathbf{ZY} & \mathbf{Y} \\ \mathbf{Z} & 1 \end{bmatrix}$

3) $\begin{bmatrix} a & 0 \\ 0 & \frac{1}{a} \end{bmatrix} \begin{bmatrix} 1 + \mathbf{ZY} & \mathbf{Z} \\ \mathbf{Y} & 1 \end{bmatrix}$

4) $\begin{bmatrix} \frac{1}{a} & 0 \\ 0 & a \end{bmatrix} \begin{bmatrix} 1 + \mathbf{ZY} & \mathbf{Y} \\ \mathbf{Z} & 1 \end{bmatrix}$

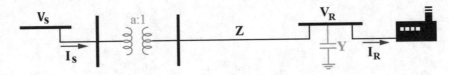

Fig. 5.4 The power system of problem 5.16

Solutions of Problems: Transmission Line Model and Performance

Abstract

In this chapter, the problems of the fifth chapter are fully solved, in detail, step by step, and with different methods.

6.1. In a short transmission line, the capacitance of the line can be ignored. Choice (4) is the answer.

6.2. Based on Ferranti effect, the voltage in the receiving end increases when the transmission line is operated in no-load or low-load conditions. Choice (1) is the answer.

6.3. The two-port of transmission line is symmetric and bidirectional. Therefore, the transmission matrix of a real transmission line has the following characteristics:

$$[T] = \begin{bmatrix} A & B \\ C & D \end{bmatrix} \Rightarrow \begin{cases} A = D & (1) \\ \det([T]) = 1 \Rightarrow AD - BC = 1 & (2) \end{cases}$$

Now, we need to check these characteristics for each choice, as follows:

Choice 1: Condition (1) is not true, as $j \neq (-j)$.
Choice 2: Condition (2) is not true, as $AD - BC = 1 - j2 \neq 1$.
Choice 3: Condition (2) is not true, as $AD - BC = 1 - 6 \neq 1$.
Choice 4: Both conditions are true, as $1 = 1$ and $AD - BC = 1 - 0 = 1$.

Therefore, the transmission matrix of choice 4 belongs to a real transmission line. Choice (4) is the answer.

6.4. Based on the information given in the problem, we have:

$$[T_1] = \begin{bmatrix} 1 & j2 \\ 0 & 1 \end{bmatrix}, [T_2] = \begin{bmatrix} 1 & 0 \\ j2 & 1 \end{bmatrix} \tag{1}$$

As we know, for the cascaded transmission systems, the relation below holds about their transmission matrices:

$$[T_{Total}] = [T_1] \times [T_2] \tag{2}$$

Solving (1) and (2):

$$[T_{Total}] = \begin{bmatrix} 1 & j2 \\ 0 & 1 \end{bmatrix} \times \begin{bmatrix} 1 & 0 \\ j2 & 1 \end{bmatrix} \Rightarrow [T_{Total}] = \begin{bmatrix} -3 & j2 \\ j2 & 1 \end{bmatrix}$$

Choice (1) is the answer.

© The Author(s), under exclusive license to Springer Nature Switzerland AG 2022
M. Rahmani-Andebili, *Power System Analysis*, https://doi.org/10.1007/978-3-030-84767-8_6

6.5. Based on the information given in the problem, we know that the line is lossless. Therefore:

$$\begin{cases} \mathbf{Z} = R + jX \xrightarrow{\;R=0,\,X=L\omega\;} \mathbf{Z} = jL\omega \\ \mathbf{Y} = G + jB \xrightarrow{\;G=0,\,B=C\omega\;} \mathbf{Y} = jC\omega \end{cases}$$

(1)

(2)

Moreover:

$$L = 1\, mH/meter, \quad C = 10\, \mu F/meter$$

(3)

As we know, the characteristic impedance of a transmission line can be calculated as follows:

$$\mathbf{Z_C} = \sqrt{\frac{\mathbf{Z}}{\mathbf{Y}}}$$

(4)

Solving (1)–(4):

$$\mathbf{Z_C} = \sqrt{\frac{\mathbf{Z}}{\mathbf{Y}}} = \sqrt{\frac{jL\omega}{jC\omega}} = \sqrt{\frac{L}{C}} = \sqrt{\frac{10^{-3}}{10 \times 10^{-6}}} = \sqrt{\frac{1}{10^{-2}}}$$

$$\mathbf{Z_C} = 10\, \Omega$$

Choice (3) is the answer.

6.6. Based on the information given in the problem, we have:

$$\mathbf{Z_C} = (1 - j)\, \Omega$$

(1)

$$\mathbf{Z_L} = (1 + j)\, \Omega$$

(2)

If a transmission line is loaded by an impedance, which is equal to its characteristic impedance, the reflected waves of the voltage and current will be eliminated. Therefore, we need an impedance ($\mathbf{Z}$) to install it parallel to the load impedance ($\mathbf{Z_L}$) to achieve the goal ($\mathbf{Z_C}$):

$$\mathbf{Z_L} \| \mathbf{Z} = \mathbf{Z_C}$$

$$\Rightarrow (1+j)\|\mathbf{Z} = 1-j \Rightarrow \frac{(1+j) \times \mathbf{Z}}{(1+j) + \mathbf{Z}} = 1 - j$$

$$\Rightarrow \mathbf{Z} + j\mathbf{Z} = (1+j) + \mathbf{Z} - j + 1 - j\mathbf{Z} \Rightarrow 2j\mathbf{Z} = 2$$

$$\Rightarrow \mathbf{Z} = \frac{1}{j}\, \Omega \Rightarrow X_C = 1\, \Omega$$

Therefore, the component is a capacitor with the reactance of $1\, \Omega$. Choice (2) is the answer.

6.7. As we know, the transmission matrix of the T model of a medium transmission line can be presented as follows:

$$\begin{bmatrix} \mathbf{V_S} \\ \mathbf{I_S} \end{bmatrix} = \begin{bmatrix} 1 + \dfrac{\mathbf{ZY}}{2} & \mathbf{Z}\left(1 + \dfrac{\mathbf{ZY}}{4}\right) \\ \mathbf{Y} & 1 + \dfrac{\mathbf{ZY}}{2} \end{bmatrix} \begin{bmatrix} \mathbf{V_R} \\ \mathbf{I_R} \end{bmatrix} \Rightarrow \begin{cases} \mathbf{V_S} = \left(1 + \dfrac{\mathbf{ZY}}{2}\right)\mathbf{V_R} + \mathbf{Z}\left(1 + \dfrac{\mathbf{ZY}}{4}\right)\mathbf{I_R} \\ \mathbf{I_S} = \mathbf{YV_R} + \left(1 + \dfrac{\mathbf{ZY}}{2}\right)\mathbf{I_R} \end{cases}$$

(1)

The charging current ($I_{Charging}$) of a transmission line is achieved when the line is in the no-load condition. In other words:

$$I_R = 0, \quad I_S = I_{Charging} \tag{2}$$

Solving (1) and (2):

$$V_S = \left(1 + \frac{ZY}{2}\right)V_R \tag{3}$$

$$I_{Charging} = YV_R \tag{4}$$

Solving (3) and (4):

$$I_{Charging} = YV_S\left(1 + \frac{ZY}{2}\right)^{-1}$$

Choice (3) is the answer.

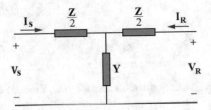

Fig. 6.1 The power system of solution of problem 6.7

6.8. As we know, transmission matrix is presented in the following form:

$$\begin{bmatrix} V_S \\ I_S \end{bmatrix} = \begin{bmatrix} A & B \\ C & D \end{bmatrix} \begin{bmatrix} V_R \\ I_R \end{bmatrix} \tag{1}$$

Applying KVL:

$$V_S = ZI_S + V_R \tag{2}$$

Applying KCL in the receiving end:

$$-I_S + V_RY + I_R = 0 \Rightarrow I_S = V_RY + I_R \tag{3}$$

Solving (2) and (3):

$$V_S = Z(V_RY + I_R) + V_R = V_R(ZY + 1) + ZI_R \tag{4}$$

Solving (1), (3), and (4):

$$\begin{bmatrix} V_S \\ I_S \end{bmatrix} = \begin{bmatrix} 1 + ZY & Z \\ Y & 1 \end{bmatrix} \begin{bmatrix} V_R \\ I_R \end{bmatrix}$$

Choice (2) is the answer.

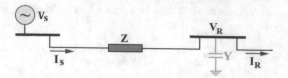

Fig. 6.2 The power system of solution of problem 6.8

6.9. Based on the information given in the problem, we have:

$$\frac{R}{L} = \frac{G}{C} \tag{1}$$

As we know, the characteristic impedance of a transmission line can be calculated as follows:

$$\mathbf{Z_C} = \sqrt{\frac{\mathbf{Z}}{\mathbf{Y}}} = \sqrt{\frac{R + j\omega L}{G + j\omega C}} = \sqrt{\frac{L}{C}\left(\frac{\frac{R}{L} + j\omega}{\frac{G}{C} + j\omega}\right)} \tag{2}$$

Solving (1) and (2):

$$\mathbf{Z_C} = \sqrt{\frac{L}{C}} \tag{3}$$

The characteristic impedance of a lossless transmission line ($R = 0$, $G = 0$) can be determined as follows:

$$\mathbf{Z_C} = \sqrt{\frac{\mathbf{Z}}{\mathbf{Y}}} = \sqrt{\frac{0 + j\omega L}{0 + j\omega C}} = \sqrt{\frac{j\omega L}{j\omega C}} = \sqrt{\frac{L}{C}} \tag{4}$$

By comparing (3) and (4), it is concluded that choice (4) is the answer.

6.10. As we know, the transmission matrix is presented as follows:

$$\begin{bmatrix} \mathbf{V_S} \\ \mathbf{I_S} \end{bmatrix} = \begin{bmatrix} A & B \\ C & D \end{bmatrix} \begin{bmatrix} \mathbf{V_R} \\ \mathbf{I_R} \end{bmatrix} \tag{1}$$

This power system should be considered as the three cascaded sub-systems, as is illustrated in Fig. 6.3.2–4. Then, the relation below holds about their transmission matrices:

$$[T_{Total}] = [T_1] \times [T_2] \times [T_3] \tag{2}$$

Note that since the resistor has been installed in the middle point, the impedance of the line ($\mathbf{Z}$) is equally divided.

The transmission matrix of the first or the third sub-system (see Fig. 6.3.2 and Fig. 6.3.4) can be determined as follows:

Applying KVL:

$$\mathbf{V_S} = \frac{1}{2}\mathbf{Z}\mathbf{I_{R1}} + \mathbf{V_{R1}} \tag{3}$$

Applying KCL:

$$\mathbf{I_S} = \mathbf{I_{R1}} \tag{4}$$

Solving (1), (3), and (4):

$$\begin{bmatrix} \mathbf{V_S} \\ \mathbf{I_S} \end{bmatrix} = \begin{bmatrix} 1 & \frac{1}{2}\mathbf{Z} \\ 0 & 1 \end{bmatrix} \begin{bmatrix} \mathbf{V_{R1}} \\ \mathbf{I_{R1}} \end{bmatrix} \Rightarrow [T_1] = [T_3] = \begin{bmatrix} 1 & \frac{1}{2}\mathbf{Z} \\ 0 & 1 \end{bmatrix} \tag{5}$$

The transmission matrix of the second sub-system (see Fig. 6.3.3) can be determined as follows:

Applying KVL:

$$\mathbf{V_{S2}} = \mathbf{V_{R2}} \tag{6}$$

Applying KCL:

$$-\mathbf{I_{S2}} + \frac{\mathbf{V_{R2}}}{R} + \mathbf{I_{R2}} = 0 \Rightarrow \mathbf{I_{S2}} = \frac{1}{R}\mathbf{V_{R2}} + \mathbf{I_{R2}} \tag{7}$$

Solving (1), (6), and (7):

$$\begin{bmatrix} \mathbf{V_{S2}} \\ \mathbf{I_{S2}} \end{bmatrix} = \begin{bmatrix} 1 & 0 \\ \frac{1}{R} & 1 \end{bmatrix} \begin{bmatrix} \mathbf{V_{R2}} \\ \mathbf{I_{R2}} \end{bmatrix} \Rightarrow [T_2] = \begin{bmatrix} 1 & 0 \\ \frac{1}{R} & 1 \end{bmatrix} \tag{8}$$

Solving (2), (5), and (8):

$$[T_{Total}] = \begin{bmatrix} 1 & \frac{1}{2}\mathbf{Z} \\ 0 & 1 \end{bmatrix} \times \begin{bmatrix} 1 & 0 \\ \frac{1}{R} & 1 \end{bmatrix} \times \begin{bmatrix} 1 & \frac{1}{2}\mathbf{Z} \\ 0 & 1 \end{bmatrix}$$

$$\Rightarrow [T_{Total}] = \begin{bmatrix} 1 + \frac{Z}{2R} & Z\left(R + \frac{Z}{4R}\right) \\ \frac{1}{R} & 1 + \frac{Z}{2R} \end{bmatrix}$$

Choice (3) is the answer.

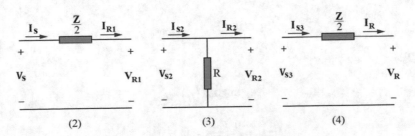

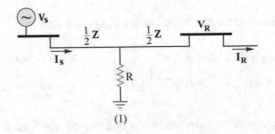

Fig. 6.3 The power system of solution of problem 6.10

6.11. Based on the information given in the problem, we have:

$$[T] = \begin{bmatrix} \dfrac{1}{2} & j \\ \dfrac{3}{4}j & \dfrac{1}{2} \end{bmatrix} \tag{1}$$

As we know, the transmission matrix of a long transmission line is as follows:

$$\begin{bmatrix} \mathbf{V_S} \\ \mathbf{I_S} \end{bmatrix} = \begin{bmatrix} A & B \\ C & D \end{bmatrix} \begin{bmatrix} \mathbf{V_R} \\ \mathbf{I_R} \end{bmatrix} \tag{2}$$

$$\begin{bmatrix} \mathbf{V_S} \\ \mathbf{I_S} \end{bmatrix} = \begin{bmatrix} \cosh(\gamma l) & \mathbf{Z_C}\sinh(\gamma l) \\ \dfrac{1}{\mathbf{Z_C}}\sinh(\gamma l) & \cosh(\gamma l) \end{bmatrix} \begin{bmatrix} \mathbf{V_R} \\ \mathbf{I_R} \end{bmatrix} \tag{3}$$

By considering (2) and (3), we can write:

$$\sqrt{\frac{AB}{CD}} = \sqrt{\frac{\cosh(\gamma l) \times \mathbf{Z_C}\sinh(\gamma l)}{\frac{1}{\mathbf{Z_C}}\sinh(\gamma l) \times \cosh(\gamma l)}} = \sqrt{\frac{\mathbf{Z_C}}{\frac{1}{\mathbf{Z_C}}}} = \mathbf{Z_C}$$

Moreover, as we know, $A = D$ in the transmission matrix of a transmission line. Therefore, the characteristic impedance of a transmission line can be determined as follows:

$$\mathbf{Z_C} = \sqrt{\frac{B}{C}} \tag{4}$$

Solving (1) and (4):

$$\mathbf{Z_C} = \sqrt{\frac{j}{\frac{3}{4}j}} \Rightarrow \mathbf{Z_C} = \frac{2\sqrt{3}}{3}\,\Omega$$

Choice (2) is the answer.

6.12. As we know, the transmission matrix of a transmission line is as follows:

$$\begin{bmatrix} \mathbf{V_S} \\ \mathbf{I_S} \end{bmatrix} = \begin{bmatrix} \cosh(\gamma l) & \mathbf{Z_C}\sinh(\gamma l) \\ \dfrac{1}{\mathbf{Z_C}}\sinh(\gamma l) & \cosh(\gamma l) \end{bmatrix} \begin{bmatrix} \mathbf{V_R} \\ \mathbf{I_R} \end{bmatrix} \Rightarrow \begin{cases} \mathbf{V_S} = \cosh(\gamma l)\mathbf{V_R} + \mathbf{Z_C}\sinh(\gamma l)\mathbf{I_R} \\ \mathbf{I_S} = \dfrac{1}{\mathbf{Z_C}}\sinh(\gamma l)\mathbf{V_R} + \cosh(\gamma l)\mathbf{I_R} \end{cases} \tag{1}$$

The charging current ($\mathbf{I_{Charging}}$) of a transmission line is achieved when the line is in the no-load condition. In other words:

$$\mathbf{I_R} = 0, \quad \mathbf{I_S} = \mathbf{I_{Charging}} \tag{2}$$

Solving (1) and (2):

$$\mathbf{V_S} = \cosh(\gamma l)\mathbf{V_R} \tag{3}$$

$$\mathbf{I_{Charging}} = \frac{1}{\mathbf{Z_C}}\sinh(\gamma l)\mathbf{V_R} \tag{4}$$

Solving (3) and (4):

$$\frac{V_S}{I_{Charging}} = \frac{\cosh{(\gamma l)}}{\frac{1}{Z_C}\sinh{(\gamma l)}} \Rightarrow I_{Charging} = \frac{1}{Z_C}\tanh{(\gamma l)}V_S$$

Choice (1) is the answer.

6.13. Based on the information given in the problem, we have:

Z_C: Characteristic impedance
$Z_{S.C.}$: The impedance seen from the beginning of the transmission line if its end is short circuit
$Z_{O.C.}$: The impedance seen from the beginning of the transmission line if its end is open circuit

As we know, the transmission matrix of a transmission line is as follows:

$$\begin{bmatrix} V_S \\ I_S \end{bmatrix} = \begin{bmatrix} \cosh{(\gamma l)} & Z_C\sinh{(\gamma l)} \\ \frac{1}{Z_C}\sinh{(\gamma l)} & \cosh{(\gamma l)} \end{bmatrix}\begin{bmatrix} V_R \\ I_R \end{bmatrix} \Rightarrow \begin{cases} V_S = \cosh{(\gamma l)}V_R + Z_C\sinh{(\gamma l)}I_R \\ I_S = \frac{1}{Z_C}\sinh{(\gamma l)}V_R + \cosh{(\gamma l)}I_R \end{cases} \tag{1}$$

Therefore:

$$Z_{S.C.} = \left.\frac{V_S}{I_S}\right|_{V_R=0} = \frac{Z_C\sinh{(\gamma l)}I_R}{\cosh{(\gamma l)}I_R} = Z_C\tanh{(\gamma l)} \tag{2}$$

$$Z_{O.C.} = \left.\frac{V_S}{I_S}\right|_{I_R=0} = \frac{\cosh{(\gamma l)}V_R}{\frac{1}{Z_C}\sinh{(\gamma l)}V_R} = Z_C\coth{(\gamma l)} \tag{3}$$

By considering (2) and (3), we can write:

$$Z_{S.C.}Z_{O.C.} = Z_C\tanh{(\gamma l)} \times Z_C\coth{(\gamma l)} \tag{4}$$

From trigonometry, we know that:

$$\tanh{(\gamma l)}\coth{(\gamma l)} = 1 \tag{5}$$

Solving (4) and (5):

$$Z_{S.C.}Z_{O.C.} = (Z_C)^2 \Rightarrow Z_C = \sqrt{Z_{S.C.}Z_{O.C.}}$$

Choice (3) is the answer.

6.14. Based on the information given in the problem, we have:

$$Z_{O.C.} = \frac{1}{Z_{S.C.}} \tag{1}$$

As we know, the transmission matrix is presented as follows:

$$\begin{bmatrix} V_S \\ I_S \end{bmatrix} = \begin{bmatrix} A & B \\ C & D \end{bmatrix}\begin{bmatrix} V_R \\ I_R \end{bmatrix} \tag{2}$$

Therefore:

$$\mathbf{Z}_{S.C.} = \frac{\mathbf{V_S}}{\mathbf{I_S}}\bigg|_{\mathbf{V_R}=0} = \frac{B\mathbf{I_R}}{D\mathbf{I_R}} = \frac{B}{D} \tag{3}$$

$$\mathbf{Z}_{O.C.} = \frac{\mathbf{V_S}}{\mathbf{I_S}}\bigg|_{\mathbf{I_R}=0} = \frac{A\mathbf{V_R}}{C\mathbf{V_R}} = \frac{A}{C} \tag{4}$$

Solving (1), (3), and (4):

$$\frac{A}{C} = \frac{1}{\frac{B}{D}} \Rightarrow \frac{A}{C} = \frac{D}{B} \tag{5}$$

Since the two-port of transmission line is symmetric and bidirectional, its transmission matrix has the following features:

$$A = D \tag{6}$$

$$AD - BC = 1 \tag{7}$$

Solving (5)–(7):

$$A^2 - B^2 = 1 \Rightarrow (A+B)(A-B) = 1 \Rightarrow A + B = \frac{1}{A-B}$$

Choice (1) is the answer.

6.15. As we know, the transmission matrix of a transmission line is as follows:

$$\begin{bmatrix} \mathbf{V_S} \\ \mathbf{I_S} \end{bmatrix} = \begin{bmatrix} \cosh(\gamma l) & \mathbf{Z_C}\sinh(\gamma l) \\ \frac{1}{\mathbf{Z_C}}\sinh(\gamma l) & \cosh(\gamma l) \end{bmatrix} \begin{bmatrix} \mathbf{V_R} \\ \mathbf{I_R} \end{bmatrix} \Rightarrow \begin{cases} \mathbf{V_S} = \cosh(\gamma l)\mathbf{V_R} + \mathbf{Z_C}\sinh(\gamma l)\mathbf{I_R} & (1) \\ \mathbf{I_S} = \frac{1}{\mathbf{Z_C}}\sinh(\gamma l)\mathbf{V_R} + \cosh(\gamma l)\mathbf{I_R} & (2) \end{cases}$$

In a lossless transmission line, the attenuation coefficient is zero ($\alpha = 0$). Therefore:

$$\gamma = \alpha + j\beta = j\beta \tag{3}$$

Solving (1)–(3):

$$\Rightarrow \begin{cases} \mathbf{V_S} = \cosh(j\beta l)\mathbf{V_R} + \mathbf{Z_C}\sinh(j\beta l)\mathbf{I_R} & (4) \\ \mathbf{I_S} = \frac{1}{\mathbf{Z_C}}\sinh(j\beta l)\mathbf{V_R} + \cosh(j\beta l)\mathbf{I_R} & (5) \end{cases}$$

From trigonometry, we know that:

$$\cosh(j\beta l) = \cos(\beta l) \tag{6}$$

$$\sinh(j\beta l) = j\sin(\beta l) \tag{7}$$

Solving (4)–(7):

$$\Rightarrow \begin{cases} \mathbf{V_S} = \cos(\beta l)\mathbf{V_R} + j\mathbf{Z_C}\sin(\beta l)\mathbf{I_R} & (8) \\ \mathbf{I_S} = j\frac{1}{\mathbf{Z_C}}\sin(\beta l)\mathbf{V_R} + \cos(\beta l)\mathbf{I_R} & (9) \end{cases}$$

Moreover, in a no-load transmission line, we have:

$$\mathbf{I_R} = 0 \tag{10}$$

Solving (8)–(10):

$$\Rightarrow \begin{cases} \mathbf{V_S} = \cos\left(\beta l\right)\mathbf{V_R} & (11) \\ \mathbf{I_S} = j\dfrac{1}{\mathbf{Z_C}}\sin\left(\beta l\right)\mathbf{V_R} & (12) \end{cases}$$

From (11), we can write:

$$\mathbf{V_R} = \frac{\mathbf{V_S}}{\cos\left(\beta l\right)}$$

Choice (3) is the answer.

6.16. As we know, the transmission matrix is presented as follows:

$$\begin{bmatrix} \mathbf{V_S} \\ \mathbf{I_S} \end{bmatrix} = \begin{bmatrix} A & B \\ C & D \end{bmatrix}\begin{bmatrix} \mathbf{V_R} \\ \mathbf{I_R} \end{bmatrix} \tag{1}$$

This power system should be considered as the three cascaded sub-systems, as is shown in Fig. 6.4.2–4. Then, the total transmission matrix can be determined as follows:

$$[T_{Total}] = [T_1] \times [T_2] \times [T_3] \tag{2}$$

As we know, for the ideal transformer, shown in Fig. 6.4.2, the relations below can be written:

$$\mathbf{V_S} = a\mathbf{V_{R1}} \tag{3}$$

$$\mathbf{I_S} = \frac{1}{a}\mathbf{I_{R1}} \tag{4}$$

Thus, the transmission matrix of the first sub-system is as follows:

$$\begin{bmatrix} \mathbf{V_S} \\ \mathbf{I_S} \end{bmatrix} = \begin{bmatrix} a & 0 \\ 0 & \dfrac{1}{a} \end{bmatrix}\begin{bmatrix} \mathbf{V_{R1}} \\ \mathbf{I_{R1}} \end{bmatrix} \Rightarrow [T_1] = \begin{bmatrix} a & 0 \\ 0 & \dfrac{1}{a} \end{bmatrix} \tag{5}$$

The transmission matrix of the second sub-system can be determined as follows:

Applying KVL:

$$\mathbf{V_{S2}} = \mathbf{Z}\mathbf{I_{R2}} + \mathbf{V_{R2}} \tag{6}$$

Applying KCL:

$$\mathbf{I_{S2}} = \mathbf{I_{R2}} \tag{7}$$

Solving (1), (6), and (7):

$$\begin{bmatrix} \mathbf{V_{S2}} \\ \mathbf{I_{S2}} \end{bmatrix} = \begin{bmatrix} 1 & \mathbf{Z} \\ 0 & 1 \end{bmatrix}\begin{bmatrix} \mathbf{V_{R2}} \\ \mathbf{I_{R2}} \end{bmatrix} \Rightarrow [T_2] = \begin{bmatrix} 1 & \mathbf{Z} \\ 0 & 1 \end{bmatrix} \tag{8}$$

The transmission matrix of the third sub-system can be determined as follows:

Applying KVL:

$$V_{S3} = V_R \tag{9}$$

Applying KCL:

$$-I_{S3} + YV_R + I_R = 0 \Rightarrow I_{S3} = YV_R + I_R \tag{10}$$

Solving (1), (9), and (10):

$$\begin{bmatrix} V_{S3} \\ I_{S3} \end{bmatrix} = \begin{bmatrix} 1 & 0 \\ Y & 1 \end{bmatrix} \begin{bmatrix} V_R \\ I_R \end{bmatrix} \Rightarrow [T_3] = \begin{bmatrix} 1 & 0 \\ Y & 1 \end{bmatrix} \tag{11}$$

Solving (2), (5), (8), and (11):

$$[T_{Total}] = \begin{bmatrix} a & 0 \\ 0 & \dfrac{1}{a} \end{bmatrix} \times \begin{bmatrix} 1 & Z \\ 0 & 1 \end{bmatrix} \times \begin{bmatrix} 1 & 0 \\ Y & 1 \end{bmatrix}$$

$$[T_{Total}] = \begin{bmatrix} \dfrac{1}{a} & 0 \\ 0 & a \end{bmatrix} \begin{bmatrix} 1 + ZY & Y \\ Z & 1 \end{bmatrix}$$

Choice (3) is the answer.

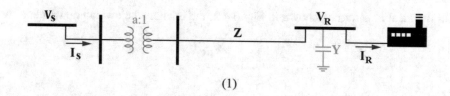

(1)

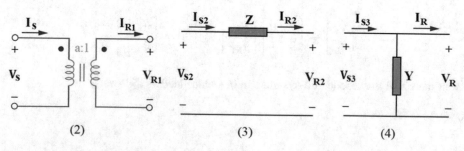

(2)   (3)   (4)

Fig. 6.4 The power system of solution of problem 6.16

Abstract

In this chapter, the problems of network impedance and admittance matrices are presented. In this chapter, the problems are categorized in different levels based on their difficulty levels (easy, normal, and hard) and calculation amounts (small, normal, and large). Additionally, the problems are ordered from the easiest problem with the smallest computations to the most difficult problems with the largest calculations.

7.1. For the power system illustrated in Fig. 7.1, determine Z_{22} of the network impedance matrix ($[Z_{Bus}]$).

Difficulty level   ○ Easy   ● Normal   ○ Hard
Calculation amount   ● Small   ○ Normal   ○ Large

1) $j0.6 \, \Omega$
2) $j0.06 \, \Omega$
3) $j0.4 \, \Omega$
4) $j0.15 \, \Omega$

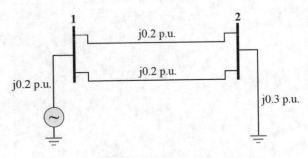

Fig. 7.1 The power system of problem 7.1

7.2. The network impedance matrix ($[Z_{Bus}]$) and the result of load flow simulation problem are presented in the following. If a capacitor with the reactance of 3.4 $p.u.$ is connected to the fourth bus, determine its updated voltage:

$$[Z_{Bus}] = j \begin{bmatrix} 0.20 & 0.15 & 0.25 & 0.24 \\ 0.15 & 0.30 & 0.13 & 0.14 \\ 0.25 & 0.13 & 0.15 & 0.25 \\ 0.24 & 0.14 & 0.25 & 0.40 \end{bmatrix} p.u.$$

Bus	1	2	3	4
V ($p.u.$)	$1.02 \angle 0°$	$0.98 \angle -15°$	$1.05 \angle 10°$	$0.9 \angle 0°$

Difficulty level   ○ Easy   ● Normal   ○ Hard
Calculation amount   ● Small   ○ Normal   ○ Large

1) $0.95\ p.\,u.$

2) $0.98\ p.\,u.$

3) $1.02\ p.\,u.$

4) $1.20\ p.\,u.$

7.3. In a three-bus power system, the voltage of the second bus is about $(1.2\ \angle\underline{0°})\ p.\,u.$, and the network impedance matrix is as follows. If an inductor with the reactance of $2.7\ p.\,u.$ is connected to the second bus, determine the voltage variation of the third bus:

$$[Z_{Bus}] = j\begin{bmatrix} 0.2 & 0.15 & 0.1 \\ 0.15 & 0.3 & 0.15 \\ 0.1 & 0.15 & 0.25 \end{bmatrix}\ p.u.$$

Difficulty level   ○ Easy   ● Normal   ○ Hard
Calculation amount   ● Small   ○ Normal   ○ Large

1) $0.075\ p.\,u.$

2) $0.06\ p.\,u.$

3) $0.12\ p.\,u.$

4) $0.15\ p.\,u.$

7.4. For the power system shown in Fig. 7.2, determine the network admittance matrix ($[Y_{Bus}]$).

Difficulty level   ○ Easy   ● Normal   ○ Hard
Calculation amount   ○ Small   ● Normal   ○ Large

1) $j\begin{bmatrix} -20 & 15 & 15 \\ 15 & -25 & 10 \\ 15 & 10 & -30 \end{bmatrix}\ p.u.$

2) $j\begin{bmatrix} -25 & 10 & 5 \\ 10 & -35 & 5 \\ 5 & 5 & -15 \end{bmatrix}\ p.u.$

3) $j\begin{bmatrix} -20 & 5 & 10 \\ 5 & -30 & 15 \\ 10 & 15 & -35 \end{bmatrix}\ p.u.$

4) $j\begin{bmatrix} -15 & 30 & -40 \\ 30 & -20 & 20 \\ -40 & 20 & -35 \end{bmatrix}\ p.u.$

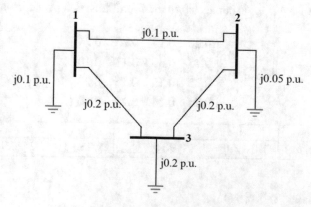

Fig. 7.2 The power system of problem 7.4

7.5. For the power system shown in Fig. 7.3, determine the network impedance matrix ($[Z_{Bus}]$).

Difficulty level ○ Easy ● Normal ○ Hard
Calculation amount ○ Small ● Normal ○ Large

1) $j \begin{bmatrix} \dfrac{2}{30} & \dfrac{1}{30} \\ \dfrac{1}{30} & \dfrac{2}{30} \end{bmatrix} p.u.$

2) $j \begin{bmatrix} \dfrac{1}{15} & -\dfrac{1}{15} \\ -\dfrac{1}{15} & \dfrac{1}{15} \end{bmatrix} p.u.$

3) $j \begin{bmatrix} \dfrac{2}{15} & \dfrac{2}{15} \\ \dfrac{2}{15} & \dfrac{2}{15} \end{bmatrix} p.u.$

4) $j \begin{bmatrix} -\dfrac{2}{30} & -\dfrac{1}{30} \\ -\dfrac{1}{30} & -\dfrac{2}{30} \end{bmatrix} p.u.$

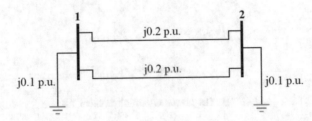

Fig. 7.3 The power system of problem 7.5

7.6. For the power system shown in Fig. 7.4, determine the detriment of the network impedance matrix ($[Z_{Bus}]$).

Difficulty level ○ Easy ● Normal ○ Hard
Calculation amount ○ Small ● Normal ○ Large

1) -0.5
2) 0.5
3) -0.2
4) 0.2

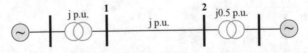

Fig. 7.4 The power system of problem 7.6

7.7. For the power system shown in Fig. 7.5, determine the value of $\frac{Z_{12}}{Z_{22}}$, belonging to $[Z_{Bus}]$, if the base voltage in the transmission line and the base MVA are 50 kV and 100 MVA, respectively.

Difficulty level ○ Easy ● Normal ○ Hard
Calculation amount ○ Small ● Normal ○ Large

1) 0.5
2) 0.75
3) 1
4) 2

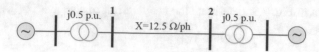

Fig. 7.5 The power system of problem 7.7

7.8. In a three-bus power system shown in Fig. 7.6, determine the sum of the diagonal components of the network admittance matrix ([Y_{Bus}]).

Difficulty level　　　○ Easy　　● Normal　　○ Hard
Calculation amount　○ Small　　● Normal　　○ Large

1) $-j60\ p.\,u.$
2) $-j20\ p.\,u.$
3) $-j30\ p.\,u.$
4) $-j10\ p.\,u.$

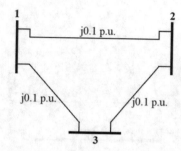

Fig. 7.6 The power system of problem 7.8

7.9. The impedance diagram of a three-phase four-bus power system is shown in Fig. 7.7. If the lines of 2–4 and 1–3 are removed from the system, the network admittance matrix can be presented in the form of [$Y_{Bus,\ New}$] = [Y_{Bus}] + [ΔY_{Bus}]. Determine [ΔY_{Bus}].

Difficulty level　　　○ Easy　　● Normal　　○ Hard
Calculation amount　○ Small　　○ Normal　　● Large

1) $j\begin{bmatrix} 10 & 0 & 10 & 0 \\ 0 & 10 & 0 & 10 \\ 10 & 0 & 10 & 0 \\ 0 & 10 & 0 & 10 \end{bmatrix} p.u.$

2) $j\begin{bmatrix} 10 & 0 & -10 & 0 \\ 0 & 10 & 0 & -10 \\ -10 & 0 & 10 & 0 \\ 0 & -10 & 0 & 10 \end{bmatrix} p.u.$

3) $j\begin{bmatrix} -10 & 0 & 10 & 0 \\ 0 & -10 & 0 & 10 \\ 10 & 0 & -10 & 0 \\ 0 & 10 & 0 & -10 \end{bmatrix} p.u.$

4) $j\begin{bmatrix} -10 & 0 & -10 & 0 \\ 0 & -10 & 0 & -10 \\ -10 & 0 & -10 & 0 \\ 0 & -10 & 0 & -10 \end{bmatrix} p.u.$

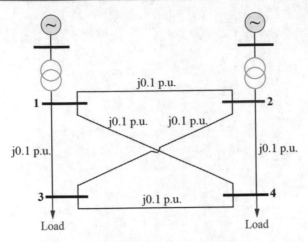

Fig. 7.7 The power system of problem 7.9

7.10. The network admittance matrix of a four-bus power system is presented in the following. Determine the updated network admittance matrix if the second and the third buses are short-circuited:

$$[Y_{Bus}] = j \begin{bmatrix} -5 & 4 & 3 & 2 \\ 4 & -10 & 2 & 1 \\ 3 & 2 & -10 & 4 \\ 2 & 1 & 4 & -20 \end{bmatrix} p.u.$$

Difficulty level   ○ Easy   ○ Normal   ● Hard
Calculation amount ● Small   ○ Normal   ○ Large

1) $j \begin{bmatrix} -5 & 7 & 2 \\ 7 & -16 & 5 \\ 2 & 5 & -20 \end{bmatrix} p.u.$

2) $j \begin{bmatrix} -10 & 7 & 2 \\ 7 & -10 & 5 \\ 2 & 5 & -10 \end{bmatrix} p.u.$

3) $j \begin{bmatrix} -16 & -5 & 6 \\ -5 & -10 & 5 \\ 6 & 5 & -20 \end{bmatrix} p.u.$

4) $j \begin{bmatrix} 0 & 10 & 20 \\ 10 & -10 & 6 \\ 20 & 6 & -20 \end{bmatrix} p.u.$

7.11. The network admittance matrix of a power system is presented in the following. There are two parallel similar lines between the buses. If one of them is disconnected from bus 1 and then grounded, determine the updated network admittance matrix:

$$[Y_{Bus}] = \begin{bmatrix} -j10 & j10 \\ j10 & -j10 \end{bmatrix} p.u.$$

Difficulty level   ○ Easy   ○ Normal   ● Hard

Calculation amount   ○ Small   ● Normal   ○ Large

1) $j\begin{bmatrix} -5 & 5 \\ 5 & -10 \end{bmatrix} p.u.$

2) $j\begin{bmatrix} -20 & 20 \\ 20 & -20 \end{bmatrix} p.u.$

3) $j\begin{bmatrix} -20 & 5 \\ 5 & -10 \end{bmatrix} p.u.$

4) $j\begin{bmatrix} -5 & 5 \\ 5 & -5 \end{bmatrix} p.u.$

Solutions of Problems: Network Impedance and Admittance Matrices

8

Abstract

In this chapter, the problems of the seventh chapter are fully solved, in detail, step by step, and with different methods.

8.1. As we know, Z_{nn} is the Thevenin impedance seen from the $n'th$ bus. To find the Thevenin impedance, we need to turn off the generator, as is shown in Fig. 8.2. Now, we can write:

$$Z_{22} = (\, j0.3)\|((\, j0.2)\|(\, j0.2) + j0.2) = (\, j0.3)\|(\, j0.3)$$

$$Z_{22} = j0.15 \; p.u.$$

Choice (4) is the answer.

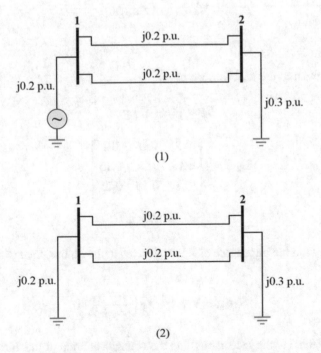

Fig. 8.1 The power system of solution of problem 8.4

8.2. Based on the information given in the problem, we have:

$$[Z_{Bus}] = j \begin{bmatrix} 0.20 & 0.15 & 0.25 & 0.24 \\ 0.15 & 0.30 & 0.13 & 0.14 \\ 0.25 & 0.13 & 0.15 & 0.25 \\ 0.24 & 0.14 & 0.25 & 0.40 \end{bmatrix} p.u.$$

Bus	1	2	3	4
$V\ (p.u.)$	$1.02\underline{/0°}$	$0.98\underline{/-15°}$	$1.05\underline{/10°}$	$0.9\underline{/0°}$

$$Z_C = -j3.4\ p.u.$$

If an inductor or a capacitor with the impedance of Z is connected to the ith bus, the updated voltage in the jth bus can be calculated as follows:

$$V_{j,New} = V_{j,Old} + Z_{ji}\left(\frac{-V_{i,Old}}{Z_{ii}+Z}\right)$$

In this problem, the capacitor is connected to the fourth bus, and the updated voltage of the fourth bus is also requested. Thus, using the network impedance matrix and the result of load flow simulation problem, we can write:

$$V_{4,New} = V_{4,Old} + Z_{44}\left(\frac{-V_{4,Old}}{Z_{44}+Z}\right) \Rightarrow V_{4,New} = 0.9\underline{/0°} + (j0.4)\left(\frac{-0.9\underline{/0°}}{j0.4+(-j3.4)}\right)$$
$$= 0.9 + \frac{-j0.36}{-j3}$$

$$V_{4,New} = 1.02\ p.u.$$

Choice (3) is the answer.

8.3. Based on the information given in the problem, we have:

$$V_2 = (1.2\underline{/0°})\ p.u.$$

$$[Z_{Bus}] = j \begin{bmatrix} 0.2 & 0.15 & 0.1 \\ 0.15 & 0.3 & 0.15 \\ 0.1 & 0.15 & 0.25 \end{bmatrix} p.u.$$

$$X_L = 2.7\ p.u.$$

If an inductor or a capacitor with the impedance of Z is connected to the ith bus, the updated voltage in the jth bus can be calculated as follows:

$$V_{j,New} = V_{j,Old} + Z_{ji}\left(\frac{-V_{i,Old}}{Z_{ii}+Z}\right)$$

Herein, the inductor is connected to the second bus, and the voltage variation of the third bus is requested. Thus, we can write:

$$\Delta V_3 = V_{3,New} - V_{3,Old} = Z_{32}\left(\frac{-V_{2,Old}}{Z_{22}+Z}\right) = (j0.15)\left(\frac{-1.2\underline{/0°}}{j0.3+j2.7}\right) \Rightarrow |\Delta V_3| = 0.06\ p.u.$$

Choice (2) is the answer.

8.4. Figure 8.2 shows the power system. The components of the network admittance matrix ($[Y_{Bus}]$) can be determined as follows:

$$y_{11} = \frac{1}{j0.1} + \frac{1}{j0.2} + \frac{1}{j0.1} = -j10 - j5 - j10 = -j25 \; p.u.$$

$$y_{22} = \frac{1}{j0.05} + \frac{1}{j0.1} + \frac{1}{j0.2} = -j20 - j10 - j5 = -j35 \; p.u.$$

$$y_{33} = \frac{1}{j0.2} + \frac{1}{j0.2} + \frac{1}{j0.2} = -j5 - j5 - j5 = -j15 \; p.u.$$

$$y_{12} = y_{21} = -\left(\frac{1}{j0.1}\right) = j10 \; p.u.$$

$$y_{13} = y_{31} = -\left(\frac{1}{j0.2}\right) = j5 \; p.u.$$

$$y_{23} = y_{32} = -\left(\frac{1}{j0.2}\right) = j5 \; p.u.$$

Therefore:

$$[Y_{Bus}] = j \begin{bmatrix} -25 & 10 & 5 \\ 10 & -35 & 5 \\ 5 & 5 & -15 \end{bmatrix} p.u.$$

Choice (2) is the answer.

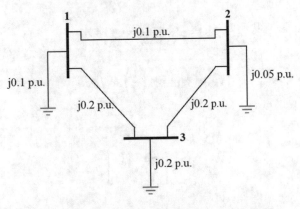

Fig. 8.2 The power system of solution of problem 8.1

8.5. Building network impedance matrix ($[Z_{Bus}]$) is time-consuming. Hence, the best way is to determine the network admittance matrix ($[Y_{Bus}]$), and then $[Z_{Bus}] = [Y_{Bus}]^{-1}$:

$$y_{11} = \frac{1}{j0.1} + \frac{1}{j0.2} + \frac{1}{j0.2} = -j10 - j5 - j5 = -j20 \; p.u.$$

$$y_{22} = \frac{1}{j0.1} + \frac{1}{j0.2} + \frac{1}{j0.2} = -j10 - j5 - j5 = -j20 \; p.u.$$

$$y_{12} = y_{21} = -\left(\frac{1}{j0.2} + \frac{1}{j0.2}\right) = -(-j5 - j5) = j10 \ p.u.$$

Therefore:

$$[Y_{Bus}] = j\begin{bmatrix} -20 & 10 \\ 10 & -20 \end{bmatrix} p.u.$$

$$[Z_{Bus}] = [Y_{Bus}]^{-1} = \left(j\begin{bmatrix} -20 & 10 \\ 10 & -20 \end{bmatrix}\right)^{-1} \Rightarrow [Z_{Bus}] = j\begin{bmatrix} \dfrac{2}{30} & \dfrac{1}{30} \\ \dfrac{1}{30} & \dfrac{2}{30} \end{bmatrix}$$

Choice (1) is the answer.

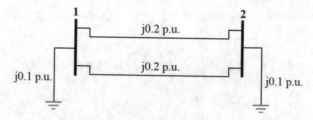

Fig. 8.3 The power system of solution of problem 8.5

8.6. Building network impedance matrix ($[Z_{Bus}]$) is time-consuming. Therefore, the best way is to determine the network admittance matrix ($[Y_{Bus}]$), and then $[Z_{Bus}] = [Y_{Bus}]^{-1}$. As is illustrated in Fig. 8.4.2, we need to turn off the generators:

$$y_{11} = \frac{1}{j} + \frac{1}{j} = -j2 \ p.u.$$

$$y_{22} = \frac{1}{j} + \frac{1}{j0.5} = -j3 \ p.u.$$

$$y_{12} = y_{21} = -\left(\frac{1}{j}\right) = j \ p.u.$$

Therefore:

$$[Y_{Bus}] = j\begin{bmatrix} -2 & 1 \\ 1 & -3 \end{bmatrix} p.u.$$

$$[Z_{Bus}] = [Y_{Bus}]^{-1} = \left(j\begin{bmatrix} -2 & 1 \\ 1 & -3 \end{bmatrix}\right)^{-1} \Rightarrow [Z_{Bus}] = j\begin{bmatrix} \dfrac{3}{5} & \dfrac{1}{5} \\ \dfrac{1}{5} & \dfrac{2}{5} \end{bmatrix}$$

$$\Rightarrow \det([Z_{Bus}]) = \left(-\frac{6}{25} - \left(-\frac{1}{25}\right)\right) = -0.2$$

Choice (3) is the answer.

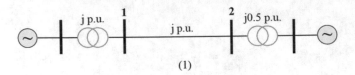

(1)

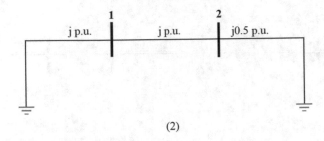

(2)

Fig. 8.4 The power system of solution of problem 8.6

8.7. Based on the information given in the problem, we have:

$$V_{Line,B} = 50 \ kV, \quad S_B = 100 \ MVA$$

$$X_{Line} = 12.5 \ \Omega$$

The base voltage in the zone of the line can be calculated as follows:

$$Z_{Line,B} = \frac{(V_{Line,B})^2}{S_B} = \frac{(50 \ kV)^2}{100 \ MVA} = 25 \ \Omega$$

Thus, the per unit (p.u.) value of the reactance of the line is:

$$X_{Line,p.u.} = \frac{X_{Line}}{Z_{Line,B}} = \frac{12.5}{25} = 0.5 \ \Omega \Rightarrow Z_{Line,p.u.} = j0.5 \ p.u.$$

Now, the impedance diagram of the system is known and illustrated in Fig. 8.5.2. The network admittance matrix of the system can be determined as follows:

$$y_{11} = y_{22} = \frac{1}{j0.5} + \frac{1}{j0.5} = -j4 \ p.u.$$

$$y_{12} = y_{21} = -\left(\frac{1}{j0.5}\right) = j2 \ p.u.$$

$$[Y_{Bus}] = j\begin{bmatrix} -4 & 2 \\ 2 & -4 \end{bmatrix} p.u.$$

Then, the network impedance matrix is:

$$[Z_{Bus}] = [Y_{Bus}]^{-1} = j\begin{bmatrix} \dfrac{1}{3} & \dfrac{1}{6} \\ \dfrac{1}{6} & \dfrac{1}{3} \end{bmatrix} p.u.$$

Therefore:

$$\frac{Z_{12}}{Z_{22}} = \frac{j\frac{1}{6}}{j\frac{1}{3}} \Rightarrow \frac{Z_{12}}{Z_{22}} = 0.5$$

Choice (1) is the answer.

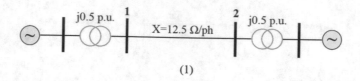

(1)

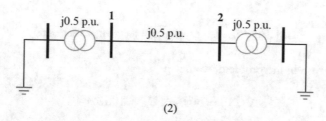

(2)

Fig. 8.5 The power system of solution of problem 8.7

8.8. The impedance diagram of the power system is shown in Fig. 8.6. The network admittance matrix of the system can be determined as follows:

$$y_{11} = y_{22} = y_{33} = \frac{1}{j0.1} + \frac{1}{j0.1} = -j10 - j10 = -j20 \; p.u.$$

$$y_{12} = y_{13} = y_{21} = y_{23} = y_{31} = y_{32} = -\left(\frac{1}{j0.1}\right) = j10 \; p.u.$$

$$[Y_{Bus}] = j \begin{bmatrix} -20 & 10 & 10 \\ 10 & -20 & 10 \\ 10 & 10 & -20 \end{bmatrix} p.u.$$

Therefore, the sum of the diagonal components of the network admittance matrix is:

$$\text{Sum of the diagonal components} = -j20 - j20 - j20 = -j60 \; p.u.$$

Choice (1) is the answer.

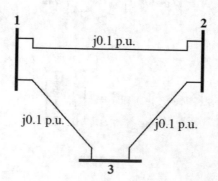

Fig. 8.6 The power system of solution of problem 8.8

8.9. Based on the information given in the problem, $[Y_{Bus}]$ belongs to the power system shown in Fig. 8.7.1. Moreover, $[Y_{Bus, New}]$ is related to the system that the lines of 2–4 and 1–3 have been removed from it.

The impedance diagram of the primary system is shown in Fig. 8.7.2. The network admittance matrix of this system can be determined as follows:

$$y_{11} = y_{22} = y_{33} = y_{44} = \frac{1}{j0.1} + \frac{1}{j0.1} + \frac{1}{j0.1} = -j10 - j10 - j10 = -j30 \, p.u.$$

$$y_{12} = y_{13} = y_{14} = y_{21} = y_{23} = y_{24} = y_{31} = y_{32} = y_{34} = y_{41} = y_{42} = y_{43} = -\left(\frac{1}{j0.1}\right) = j10 \, p.u.$$

$$[Y_{Bus}] = j \begin{bmatrix} -30 & 10 & 10 & 10 \\ 10 & -30 & 10 & 10 \\ 10 & 10 & -30 & 10 \\ 10 & 10 & 10 & -30 \end{bmatrix} p.u.$$

Figure 8.7.3 illustrates the impedance diagram of the updated system. The network admittance matrix of this system can be determined as follows:

$$y_{11} = y_{22} = y_{33} = y_{44} = \frac{1}{j0.1} + \frac{1}{j0.1} = -j10 - j10 = -j20 \, p.u.$$

$$y_{12} = y_{14} = y_{21} = y_{23} = y_{32} = y_{34} = y_{41} = y_{43} = -\left(\frac{1}{j0.1}\right) = j10 \, p.u.$$

$$y_{13} = y_{31} = y_{24} = y_{42} = 0 \, p.u.$$

$$[Y_{Bus,New}] = j \begin{bmatrix} -20 & 10 & 0 & 10 \\ 10 & -20 & 10 & 0 \\ 0 & 10 & -20 & 10 \\ 10 & 0 & 10 & -20 \end{bmatrix} p.u.$$

Therefore:

$$[\Delta Y_{Bus}] = j \begin{bmatrix} -20 & 10 & 0 & 10 \\ 10 & -20 & 10 & 0 \\ 0 & 10 & -20 & 10 \\ 10 & 0 & 10 & -20 \end{bmatrix} - j \begin{bmatrix} -30 & 10 & 10 & 10 \\ 10 & -30 & 10 & 10 \\ 10 & 10 & -30 & 10 \\ 10 & 10 & 10 & -30 \end{bmatrix} =$$

$$[\Delta Y_{Bus}] = j \begin{bmatrix} 10 & 0 & -10 & 0 \\ 0 & 10 & 0 & -10 \\ -10 & 0 & 10 & 0 \\ 0 & -10 & 0 & 10 \end{bmatrix} p.u.$$

Choice (2) is the answer.

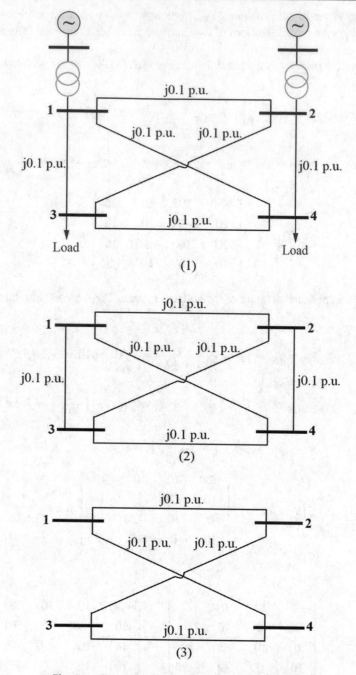

Fig. 8.7 The power system of solution of problem 8.9

8.10. By short-circuiting two buses of a power system, their corresponding components in the network admittance matrix ($[Y_{Bus}]$) are added up. Therefore, for the second and the third buses, we have:

$$[Y_{Bus}] = j \begin{bmatrix} -5 & \boxed{4 & 3} & 2 \\ \boxed{4} & -10 & 2 & \boxed{1} \\ \boxed{3} & 2 & -10 & \boxed{4} \\ 2 & \boxed{1 & 4} & -20 \end{bmatrix} \, p.u.$$

$$[Y_{Bus,New}] = j \begin{bmatrix} -5 & 7 & 2 \\ 7 & -16 & 5 \\ 2 & 5 & -20 \end{bmatrix} p.u.$$

Choice (1) is the answer.

8.11. Based on the information given in the problem, the network admittance matrix is as follows:

$$[Y_{Bus}] = \begin{bmatrix} -j10 & j10 \\ j10 & -j10 \end{bmatrix} p.u. \tag{1}$$

From this $[Y_{Bus}]$, we can figure out that the power system has only two buses.

Moreover, we know that there are two parallel similar lines between the buses. Now, it is better to draw the single-line diagram of the system which is shown in Fig. 8.8.1.

The network admittance matrix of the primary system (see Fig. 8.8.1) can be formed as follows:

$$[Y_{Bus}] = \begin{bmatrix} y+y & -(y+y) \\ -(y+y) & y+y \end{bmatrix} = \begin{bmatrix} 2y & -2y \\ -2y & 2y \end{bmatrix} p.u. \tag{2}$$

By solving (1) and (2), we can write:

$$2y = -j10 \Rightarrow y = -j5 \ p.u. \tag{3}$$

Figure 8.8.2 shows the admittance diagram of the power system. Note that each quantity presents the admittance of the line.

Based on the information given in the problem, one of them is disconnected from bus 1 and then grounded. Figure 8.8.3 illustrates the updated system. Now, the network admittance matrix of the updated system is as follows:

$$[Y_{Bus}] = \begin{bmatrix} -j5 & -(-j5) \\ -(-j5) & -j5 + (-j5) \end{bmatrix} \Rightarrow [Y_{Bus}] = \begin{bmatrix} -j5 & j5 \\ j5 & -j10 \end{bmatrix} p.u.$$

Choice (1) is the answer.

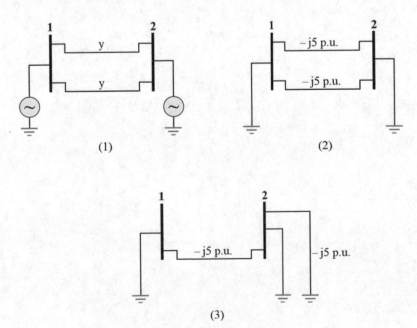

Fig. 8.8 The power system of solution of problem 8.11

Abstract

In this chapter, the problems concerned with the load flow and economic load dispatch are presented. The subjects include Gauss-Seidel load flow, DC load flow (DCLF), Decoupled Load flow (DLF), Newton-Raphson load flow (NRLF), Jacobian matrix determination, and economic load dispatch. In this chapter, the problems are categorized in different levels based on their difficulty levels (easy, normal, and hard) and calculation amounts (small, normal, and large). Additionally, the problems are ordered from the easiest problem with the smallest computations to the most difficult problems with the largest calculations.

9.1. In a load flow problem, which type of the bus has a known active power?

Difficulty level   ● Easy   ○ Normal   ○ Hard
Calculation amount   ● Small   ○ Normal   ○ Large

1) Load bus
2) Voltage-controlled bus
3) All buses except slack bus
4) None of them

9.2. To speed up the algorithm of Gauss-Seidel load flow, an accelerating factor (α) is usually used. Which one of the following relations presents that?

Difficulty level   ● Easy   ○ Normal   ○ Hard
Calculation amount   ● Small   ○ Normal   ○ Large

1) $V_{i,Acc}^{(k+1)} = V_i^{(k)} + \alpha \Delta V_i^{(k+1)}$
2) $V_{i,Acc}^{(k+1)} = \alpha V_i^{(k)} + \Delta V_i^{(k+1)}$
3) $V_{i,Acc}^{(k+1)} = \alpha \left(V_i^{(k+1)} - V_i^{(k)} \right)$
4) $V_{i,Acc}^{(k+1)} = V_i^{(k+1)} + \alpha \Delta V_i^{(k)}$

9.3. Which one of the following choices is correct about the DC load flow (DCLF), Decoupled Load flow (DLF), and Newton-Raphson load flow (NRLF)?

Difficulty level   ● Easy   ○ Normal   ○ Hard
Calculation amount   ● Small   ○ Normal   ○ Large

1) DLF is faster than DCLF, and DCLF is faster than NRLF.
2) DCLF is not appropriate for the AC power systems, and DCLF has more convergence probability compared to NRLF.
3) DLF and NRLF can achieve the same results but with different iterations. DCLF is faster than DLF and DLF is faster than NRLF.
4) DCLF is appropriate for the systems with the high value of $\frac{X}{R}$. NRLF always converges.

9.4. Use DC load flow to determine the active power flowing through the line. Herein, $S_B = 100\ MVA$.

Difficulty level   ● Easy   ○ Normal   ○ Hard
Calculation amount   ● Small   ○ Normal   ○ Large

1) 32.2 *MW*
2) 85.6 *MW*
3) 41.7 *MW*
4) 65.4 *MW*

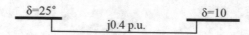

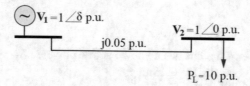

Fig. 9.1 The power system of problem 9.4

9.5. In the power system, shown in Fig. 9.2, determine δ. Do not use DC load flow approximation.

Difficulty level   ● Easy   ○ Normal   ○ Hard
Calculation amount   ● Small   ○ Normal   ○ Large

1) 60°
2) 30°
3) 90°
4) 0°

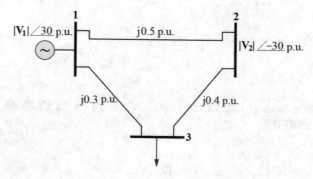

Fig. 9.2 The power system of problem 9.5

9.6. Calculate P_{12} by using DC load flow. Herein, assume $\pi \equiv 3$.

Difficulty level   ● Easy   ○ Normal   ○ Hard
Calculation amount   ● Small   ○ Normal   ○ Large

1) 1.5 *p. u.*
2) 2 *p. u.*
3) 3 *p. u.*
4) 3.5 *p. u.*

Fig. 9.3 The power system of problem 9.6

9.7. Use DC load flow to determine P_{G2}. Herein, assume $\pi \equiv 3$.

Difficulty level   ○ Easy   ● Normal   ○ Hard
Calculation amount   ○ Small   ● Normal   ○ Large

1) 0.2 *p. u.*
2) 0.25 *p. u.*
3) 0.6 *p. u.*
4) 0.75 *p. u.*

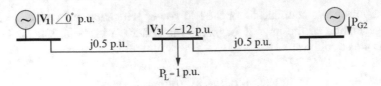

Fig. 9.4 The power system of problem 9.7

9.8. Determine the inverse matrix of Jacobian matrix considering the following terms:

$$\begin{cases} P_2 = \delta_2 + 3|\mathbf{V}_2| \\ Q_2 = 0.1\delta_2 + \dfrac{1}{5}|\mathbf{V}_1| + |\mathbf{V}_2| \end{cases}$$

Difficulty level   ○ Easy   ● Normal   ○ Hard
Calculation amount   ○ Small   ● Normal   ○ Large

1) $\begin{bmatrix} 3 & 1 \\ 1 & 0.1 \end{bmatrix}$

2) $\begin{bmatrix} 1 & 3 \\ 0.1 & 1 \end{bmatrix}$

3) $\begin{bmatrix} 0.1 & 1 \\ 1 & 3 \end{bmatrix}$

4) $\begin{bmatrix} -1 & 3 \\ 0.1 & -1 \end{bmatrix}$

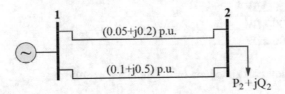

Fig. 9.5 The power system of problem 9.8

9.9. Use DC load flow to determine the phase angle of bus 4. Herein, assume $\pi \equiv 3$.

Difficulty level   ○ Easy   ● Normal   ○ Hard
Calculation amount   ○ Small   ● Normal   ○ Large

1) -45°
2) -36°
3) -30°
4) -15°

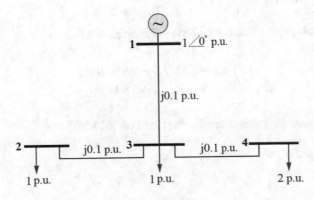

Fig. 9.6 The power system of problem 9.9

9.10. In a power plant, the power loss coefficients for the two power generation units are $L_1 = \$1.5/MW$, $L_2 = \$1.8/MW$. Calculate the total generation of the units if Lagrange Multiplier (λ) is about $\$300/MWh$, and the generation cost functions of the units are as follows:

$$\begin{cases} C_1 = 0.2P_{G1}^2 + 100P_{G1} + 5500 \\ C_2 = 0.1P_{G2}^2 + 100P_{G2} + 4000 \end{cases}$$

Difficulty level   ○ Easy   ● Normal   ○ Hard
Calculation amount   ○ Small   ● Normal   ○ Large
1) 250 MW
2) 583.3 MW
3) 425.5 MW
4) 720 MW

9.11. In a power plant, the generation cost functions of the units are as follows:

$$\begin{cases} C_1 = 0.0075P_{G1}^2 + 50P_{G1} + 1000 \\ C_2 = 0.005P_{G2}^2 + 45P_{G2} + 3000 \end{cases}$$

Solve the economic load dispatch problem for the load demand of 1000 MW.
Difficulty level   ○ Easy   ● Normal   ○ Hard
Calculation amount   ○ Small   ● Normal   ○ Large
1) $P_{G1} = 900\ MW$, $P_{G2} = 100\ MW$
2) $P_{G1} = 750\ MW$, $P_{G2} = 250\ MW$
3) $P_{G1} = 600\ MW$, $P_{G2} = 400\ MW$
4) $P_{G1} = 200\ MW$, $P_{G2} = 800\ MW$

9.12. In a power plant, the generation cost functions of the units are as follows:

$$\begin{cases} C_1 = 0.05P_{G1}^2 + 50P_{G1} + 1500 \\ C_2 = 0.075P_{G2}^2 + 40P_{G2} + 2000 \end{cases}$$

Solve the economic load dispatch problem for the total load of 1400 MW.
Difficulty level   ○ Easy   ● Normal   ○ Hard
Calculation amount   ○ Small   ● Normal   ○ Large
1) $P_{G1} = 400\ MW$, $P_{G2} = 1000\ MW$
2) $P_{G1} = 500\ MW$, $P_{G2} = 900\ MW$
3) $P_{G1} = 800\ MW$, $P_{G2} = 600\ MW$
4) $P_{G1} = 700\ MW$, $P_{G2} = 700\ MW$

9.13. In a power plant, the generation cost functions of the units are as follows:

$$\begin{cases} C_1 = 135P_{G1}^2 + 100000P_{G1} \\ C_2 = 115P_{G2}^2 + 85000P_{G2} \end{cases}$$

Solve the economic load dispatch problem for the total load of 1000 MW.
Difficulty level   ○ Easy   ● Normal   ○ Hard
Calculation amount   ○ Small   ● Normal   ○ Large

1) $P_{G1} = 430 \ MW$, $P_{G2} = 570 \ MW$
2) $P_{G1} = 570 \ MW$, $P_{G2} = 430 \ MW$
3) $P_{G1} = 500 \ MW$, $P_{G2} = 500 \ MW$
4) $P_{G1} = 536 \ MW$, $P_{G2} = 464 \ MW$

9.14. The single-line diagram of a power system is shown in Fig. 9.7. The voltage of bus 1 is about $(1 \angle 0°) \ p.u.$ and $S_B = 100 \ MVA$. Calculate $\mathbf{V_2}$ using Gauss-Seidel load flow after one iteration if $\mathbf{V}_2^{(0)} = (1 \angle 0°) \ p.u.$ and $\mathbf{V}_3^{(0)} =$

$(1 \angle 0°) \ p.u.$

Difficulty level ○ Easy ○ Normal ● Hard
Calculation amount ○ Small ○ Normal ● Large

1) $(0.936 - j0.08) \ p.u.$
2) $(0.940 - j0.08) \ p.u.$
3) $(0.8 - j0.91) \ p.u.$
4) $(0.836 - j0.2) \ p.u.$

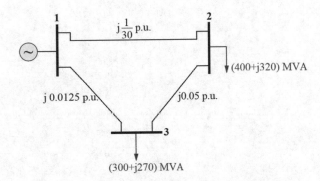

Fig. 9.7 The power system of problem 9.14

9.15. Use Newton-Raphson load flow (NRLF) to determine the voltage of load bus after one iteration.

Difficulty level ○ Easy ○ Normal ● Hard
Calculation amount ○ Small ○ Normal ● Large

1) $0.95 < -0.12 \ rad$
2) $0.98 < -0.1 \ rad$
3) $0.93 < -0.12 \ rad$
4) $0.9 < -0.1 \ rad$

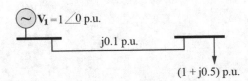

Fig. 9.8 The power system of problem 9.15

Solutions of Problems: Load Flow and Economic Load Dispatch

Abstract

In this chapter, the problems of the ninth chapter are fully solved, in detail, step by step, and with different methods.

10.1. The buses are categorized in three types:

- Load bus (P-Q bus): In this bus, the active and reactive powers are known.
- Voltage-controlled bus (P-V bus): In this bus, the active power and the magnitude of voltage are known.
- Slack bus (reference bus): In this unique bus, only the primary value of magnitude and phase angle of voltage are known.

  Therefore, active power is known in all buses except in slack bus. Choice (3) is the answer.

10.2. To speed up the algorithm of Gauss-Seidel load flow, an accelerating factor (α) is usually applied, as follows:

$$\mathbf{V}_{i,Acc}^{(k+1)} = \mathbf{V}_i^{(k)} + \alpha \Delta \mathbf{V}_i^{(k+1)}$$

Choice (1) is the answer.

10.3. DLF and NRLF can achieve the same results but with different iterations. Moreover, DC load flow (DCLF) is faster than Decoupled Load flow (DLF), and DLF is faster than Newton-Raphson load flow (NRLF). Choice (3) is the answer.

10.4. In DC load flow, the relation below is applied, in which X and δ are in per unit (p.u.) and radian, respectively:

$$P_{12} = \frac{1}{X_{12}}(\delta_1 - \delta_2) \tag{1}$$

Based on the information given in the problem, we have:

$$X_{12} = 0.4 \, p.u. \tag{2}$$

$$\delta_1 = 25°, \delta_2 = 10° \tag{3}$$

$$S_B = 100 \, MVA \tag{4}$$

Solving (1)–(3):

$$P_{12} = \frac{1}{X_{12}}(\delta_1 - \delta_2) = \frac{1}{0.4}(25 - 10) \times \frac{\pi}{180} = 0.654 \, p.u. \tag{5}$$

$$P_{MW} = P_{p.u.} \times S_B \Rightarrow P_{12} = 0.654 \times 100 = 65.4 \, MW$$

Choice (4) is the answer.

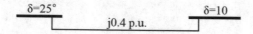

Fig. 10.1 The power system of solution of problem 10.4

10.5. Based on the information given in the problem, we have:

$$P_{12} = P_L = 10 \, p.u. \tag{1}$$

$$X_{12} = 0.05 \, p.u. \tag{2}$$

$$\delta_2 = 0° \tag{3}$$

Herein, we are not allowed to use DC load flow. The active power flowing through the transmission line can be calculated as follows:

$$P_{12} = \frac{|V_1||V_2|}{X} \sin(\delta_1 - \delta_2) \tag{1}$$

Therefore:

$$10 = \frac{1 \times 1}{0.05} \sin(\delta - 0) \Rightarrow \sin(\delta) = 0.5 \Rightarrow \delta = \sin^{-1}(0.5) = 30°$$

Choice (2) is the answer.

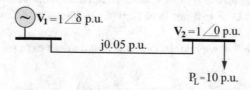

Fig. 10.2 The power system of solution of problem 10.5

10.6. In DC load flow, the relation below is applied, in which X and δ are in per unit (p.u.) and radian, respectively:

$$P_{12} = \frac{1}{X_{12}}(\delta_1 - \delta_2) \tag{1}$$

Based on the information given in the problem, we have:

$$\pi \equiv 3 \tag{2}$$

$$\delta_1 = 30° = 30 \times \frac{\pi}{180} = 0.5 \, rad \tag{3}$$

$$\delta_2 = -30° = -30 \times \frac{\pi}{180} = -0.5 \, rad \tag{4}$$

$$X_{12} = 0.5 \, p.u. \tag{5}$$

Solving (1) and (3)–(5):

$$P_{12} = \frac{1}{X_{12}} (\delta_1 - \delta_2) = \frac{1}{0.5} (0.5 - (-0.5))$$

$$P_{12} = 2 \, p.u.$$

Choice (2) is the answer.

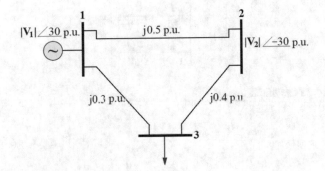

Fig. 10.3 The power system of solution of problem 10.6

10.7. Based on the information given in the problem, we have:

$$\pi \equiv 3 \tag{1}$$

$$\delta_1 = 0° = 0 \, rad \tag{2}$$

$$\delta_2 = -12° = (-12) \times \frac{\pi}{180} = -0.2 \, rad \tag{3}$$

$$X_{12} = 0.5 \, p.u. \tag{4}$$

Since there is no power loss in the lines, the total power generation will be equal to the total power demand. Hence:

$$P_{G1} + P_{G2} = P_L \Rightarrow P_{G1} + P_{G2} = 1 \, p.u. \tag{5}$$

As we know, in DC load flow, the relation below is applied, in which X and δ are in per unit (p.u.) and radian, respectively:

$$P_{12} = \frac{1}{X_{12}} (\delta_1 - \delta_2) \Rightarrow P_{G1} = P_{12} = \frac{1}{0.5} (0 - (-0.2)) = 0.4 \, p.u. \tag{6}$$

Solving (5) and (6):

$$0.4 + P_{G2} = 1 \Rightarrow P_{G2} = 0.6 \, p.u.$$

Choice (3) is the answer.

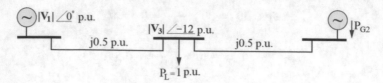

Fig. 10.4 The power system of solution of problem 10.7

10.8. Based on the information given in the problem, we have:

$$\begin{cases} P_2 = \delta_2 + 3|\mathbf{V_2}| \\ Q_2 = 0.1\delta_2 + \dfrac{1}{5}|\mathbf{V_1}| + |\mathbf{V_2}| \end{cases} \tag{1}$$

Jacobian matrix is defined as follows:

$$[J] = \begin{bmatrix} \dfrac{\partial P_2}{\partial \delta_2} & \dfrac{\partial P_2}{\partial |\mathbf{V_2}|} \\ \dfrac{\partial Q_2}{\partial \delta_2} & \dfrac{\partial Q_2}{\partial |\mathbf{V_2}|} \end{bmatrix} \tag{2}$$

Herein, bus 1 is considered as the slack bus.

Solving (1) and (2):

$$[J] = \begin{bmatrix} 1 & 3 \\ 0.1 & 1 \end{bmatrix}$$

Choice (2) is the answer.

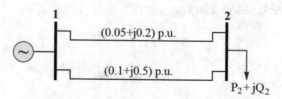

Fig. 10.5 The power system of solution of problem 10.8

10.9. In DC load flow, the relation below is applied, in which X and δ are in per unit (p.u.) and radian, respectively:

$$P_{SR} = \frac{1}{X_{SR}}(\delta_S - \delta_R) \tag{1}$$

Based on the information given in the problem, we have:

$$\pi \equiv 3 \tag{2}$$

$$\delta_1 = 0 \ rad \tag{3}$$

$$X_{13} = 0.1 \ p.u. \tag{4}$$

$$P_{13} = P_{L1} + P_{L2} + P_{L3} = 1 + 1 + 2 = 4 \ p.u. \tag{5}$$

$$P_{34} = 2 \ p.u. \tag{6}$$

Solving (1)–(5) for line 1–3:

$$P_{13} = \frac{1}{X_{13}}(\delta_1 - \delta_3) \Rightarrow 4 = \frac{1}{0.1}(0 - \delta_3) \Rightarrow \delta_3 = -0.4 \ rad \tag{7}$$

Likewise for line 3–4:

$$P_{34} = \frac{1}{X_{34}}(\delta_3 - \delta_4) \Rightarrow 2 = \frac{1}{0.1}(-0.4 - \delta_4) \Rightarrow \delta_4 = -0.6 \ rad \tag{8}$$

$$\delta_4 = -0.6 \times \frac{180}{\pi} = -36°$$

Choice (2) is the answer.

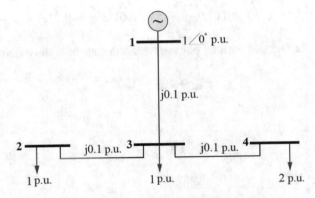

Fig. 10.6 The power system of solution of problem 10.9

10.10. Based on the information given in the problem, we have:

$$L_1 = \$1.5/MW, L_2 = \$1.8/MW \tag{1}$$

$$\lambda = \$300/MWh \tag{2}$$

$$\begin{cases} C_1 = 0.2P_{G1}^2 + 100P_{G1} + 5500 \\ C_2 = 0.1P_{G2}^2 + 100P_{G2} + 4000 \end{cases} \tag{3} \tag{4}$$

If power loss exists in a power generation system, the conditions to have an economic load dispatch are as follows:

$$\lambda = L_1 \frac{\partial C_1}{\partial P_{G1}} = L_2 \frac{\partial C_2}{\partial P_{G2}} \tag{5}$$

Solving (1)–(5):

$$300 = 1.5(0.4P_{G1} + 100) = 1.8(0.2P_{G2} + 100)$$

$$\Rightarrow \begin{cases} 300 = 1.5(0.4P_{G1} + 100) \Rightarrow 0.4P_{G1} + 100 = 200 \Rightarrow P_{G1} = 250 \ MW \\ 300 = 1.8(0.2P_{G2} + 100) \Rightarrow 0.2P_{G1} + 100 = 166.66 \Rightarrow P_{G2} = 333.3 \ MW \end{cases} \tag{6} \tag{7}$$

$$P_{G,Total} = P_{G1} + P_{G2} = 250 + 333.3 = 583.3 \ MW$$

Choice (2) is the answer.

10.11. Based on the information given in the problem, we have:

$$P_{Demand} = 1000 \, MW \tag{1}$$

$$\begin{cases} C_1 = 0.0075P_{G1}^2 + 50P_{G1} + 1000 \\ C_2 = 0.005P_{G2}^2 + 45P_{G2} + 3000 \end{cases} \tag{2}$$
$$\tag{3}$$

If the power generation system is lossless, the conditions to have an economic load dispatch are as follows:

$$\lambda = \frac{\partial C_1}{\partial P_{G1}} = \frac{\partial C_2}{\partial P_{G2}} \tag{4}$$

Solving (2)–(4):

$$0.015P_{G1} + 50 = 0.01P_{G2} + 45 \Rightarrow 0.015P_{G1} - 0.01P_{G2} = -5 \tag{5}$$

Using (1) and considering the fact that the total power generation must be equal to the total load demand, we can write:

$$P_{G1} + P_{G2} = 1000 \tag{6}$$

Solving (5) and (6):

$$P_{G1} = 200 \, MW, P_{G2} = 800 \, MW$$

Choice (4) is the answer.

10.12. Based on the information given in the problem, we have:

$$P_{Demand} = 1400 \, MW \tag{1}$$

$$\begin{cases} C_1 = 0.05P_{G1}^2 + 50P_{G1} + 1500 \\ C_2 = 0.075P_{G2}^2 + 40P_{G2} + 2000 \end{cases} \tag{2}$$
$$\tag{3}$$

If the power generation system is lossless, the conditions to have an economic load dispatch are as follows:

$$\lambda = \frac{\partial C_1}{\partial P_{G1}} = \frac{\partial C_2}{\partial P_{G2}} \tag{4}$$

Solving (2)–(4):

$$0.1P_{G1} + 50 = 0.15P_{G2} + 40 \Rightarrow 0.1P_{G1} - 0.15P_{G2} = -10 \tag{5}$$

Using (1) and considering the fact that the total power generation must be equal to the total load demand, we can write:

$$P_{G1} + P_{G2} = 1400 \tag{6}$$

Solving (5) and (6):

$$P_{G1} = 800 \, MW, P_{G2} = 600 \, MW$$

Choice (3) is the answer.

10.13. Based on the information given in the problem, we have:

$$P_{Demand} = 1000 \, MW \qquad (1)$$

$$\begin{cases} C_1 = 135P_{G1}^2 + 100000P_{G1} & (2) \\ C_2 = 115P_{G2}^2 + 85000P_{G2} & (3) \end{cases}$$

If the power generation system is lossless, the conditions to have an economic load dispatch are as follows:

$$\lambda = \frac{\partial C_1}{\partial P_{G1}} = \frac{\partial C_2}{\partial P_{G2}} \qquad (4)$$

Solving (2)–(4):

$$270P_{G1} + 100000 = 230P_{G2} + 85000 \Rightarrow 270P_{G1} - 230P_{G2} = -15000 \qquad (5)$$

Using (1) and considering the fact that the total power generation must be equal to the total load demand, we can write:

$$P_{G1} + P_{G2} = 1000 \qquad (6)$$

Solving (5) and (6):

$$P_{G1} = 430 \, MW, P_{G2} = 570 \, MW$$

Choice (1) is the answer.

10.14. Based on the information given in the problem, we have:

$$\mathbf{V_1} = (1\underline{/0°}) \, p.u. \qquad (1)$$

$$S_B = 100 \, MVA \qquad (2)$$

$$\mathbf{V_2^{(0)}} = \mathbf{V_3^{(0)}} = (1\underline{/0°}) \, p.u. \qquad (3)$$

First, we need to build the network admittance matrix, as follows:

$$y_{11} = \frac{1}{j0.0125} + \frac{1}{j\frac{1}{30}} = -j80 - j30 = -j110 \, p.u. \qquad (4)$$

$$y_{22} = \frac{1}{j0.05} + \frac{1}{j\frac{1}{30}} = -j20 - j30 = -j50 \, p.u. \qquad (5)$$

$$y_{33} = \frac{1}{j0.0125} + \frac{1}{j0.05} = -j80 - j20 = -j100 \, p.u. \qquad (6)$$

$$y_{12} = y_{21} = -\left(\frac{1}{j\frac{1}{30}}\right) = j30 \, p.u. \qquad (7)$$

$$y_{13} = y_{31} = -\left(\frac{1}{j0.0125}\right) = j80 \; p.u. \tag{8}$$

$$y_{23} = y_{32} = -\left(\frac{1}{j0.05}\right) = j20 \; p.u. \tag{9}$$

Therefore:

$$[Y_{\text{Bus}}] = j \begin{bmatrix} -110 & 30 & 80 \\ 30 & -50 & 20 \\ 80 & 20 & -100 \end{bmatrix} p.u. \tag{10}$$

Now, we need to define all the quantities in per unit (p.u.) value:

$$S_{L2,p.u.} = \frac{S_{L2}}{S_B} = \frac{400 + j320}{100} = (4 + j3.2) \; p.u. \tag{11}$$

$$S_{L3,p.u.} = \frac{S_{L3}}{S_B} = \frac{300 + j270}{100} = (3 + j2.7) \; p.u. \tag{12}$$

Based on Gauss-Seidel load flow, we have:

$$\mathbf{V}_i^{(k+1)} = \frac{1}{y_{ii}} \left(\left(\frac{P_i^{Sch} + jQ_i^{Sch}}{\mathbf{V}_i^{(k)}}\right)^* - \sum_{\substack{j=1 \\ j \neq i}}^{n} y_{ij} \mathbf{V}_j^{(k)} \right) \tag{13}$$

Where:

$$P_i^{Sch} = P_{i,G} - P_{i,L} \tag{14}$$

$$Q_i^{Sch} = Q_{i,G} - Q_{i,L} \tag{15}$$

In (14) and (15), positive and negative signs are considered for the generation power and load demand, respectively.

Now, for the second bus, we can write:

$$\mathbf{V}_2^{(1)} = \frac{1}{y_{22}} \left(\left(\frac{P_2^{Sch} + jQ_2^{Sch}}{\mathbf{V}_2^{(0)}}\right)^* - \sum_{\substack{j=1 \\ j \neq 2}}^{3} y_{2j} \mathbf{V}_j^{(0)} \right) = \frac{1}{-j50} \left(\left(\frac{-4 - j3.2}{1}\right)^* - (j30 \times 1 + j20 \times 1) \right) = \frac{-4 - j46.8}{-j50}$$

$$\mathbf{V}_2^{(1)} = (0.936 - j0.08) \; p.u.$$

Choice (1) is the answer.

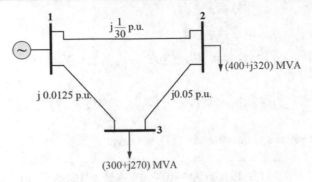

Fig. 10.7 The power system of solution of problem 10.14

10.15. Based on the information given in the problem, we have:

$$\left|V_1^{(0)}\right| = 1 \ p.u. \tag{1}$$

$$\delta_1^{(0)} = 0^\circ \tag{2}$$

$$X_{12} = 0.1 \ p.u., \quad S_{2,L} = (1 + j0.5) \ p.u. \tag{3}$$

First, we need to determine the network admittance matrix, as follows:

$$y_{11} = y_{22} = \frac{1}{j0.1} = -j10 \ p.u. \tag{4}$$

$$y_{12} = y_{21} = -\left(\frac{1}{j0.1}\right) = j10 \ p.u. \tag{5}$$

Therefore:

$$[Y_{Bus}] = j \begin{bmatrix} -10 & 10 \\ 10 & -10 \end{bmatrix} \ p.u. \tag{6}$$

The primary estimation for the magnitude of voltage and phase angle of the load bus are as follows:

$$\left|V_2^{(0)}\right| = 1 \ p.u. \tag{7}$$

$$\delta_2^{(0)} = 0^\circ \tag{8}$$

Based on Newton-Raphson load flow (NRLF), the relations below are held for a load bus:

$$P_i^{(k)} = \sum_{j=1}^{n} |V_i^{(k)}||V_j^{(k)}||y_{ij}| \cos\left(\theta_{ij} - \left(\delta_i^{(k)} - \delta_j^{(k)}\right)\right) \tag{9}$$

$$Q_i^{(k)} = -\sum_{j=1}^{n} |V_i^{(k)}||V_j^{(k)}||y_{ij}| \sin\left(\theta_{ij} - \left(\delta_i^{(k)} - \delta_j^{(k)}\right)\right) \tag{10}$$

For bus 2, we can write:

$$P_2^{(0)} = |\mathbf{V_2^{(0)}}||\mathbf{V_1^{(0)}}||y_{21}| \cos\left(\theta_{21} - \left(\delta_2^{(0)} - \delta_1^{(0)}\right)\right) + |\mathbf{V_2^{(0)}}|^2 |y_{22}| \cos\left(\theta_{22}\right) \tag{11}$$

$$Q_2^{(0)} = -|\mathbf{V_2^{(0)}}||\mathbf{V_1^{(0)}}||y_{21}| \sin\left(\theta_{21} - \left(\delta_2^{(0)} - \delta_1^{(0)}\right)\right) - |\mathbf{V_2^{(0)}}|^2 |y_{22}| \sin\left(\theta_{22}\right) \tag{12}$$

By applying the primarily estimated quantities in (11), we have:

$$P_2^{(0)} = 1 \times 1 \times 10 \cos\left(90 - (0 - 0)\right) + 1^2 \times 10 \cos\left(-90\right) = 0 \tag{13}$$

Likewise for the reactive power:

$$Q_2^{(0)} = -1 \times 1 \times 10 \sin\left(90 - (0 - 0)\right) - 1^2 \times 10 \sin\left(-90\right) = -10 + 10 = 0 \tag{14}$$

Then:

$$\Delta P_2^{(0)} = P_2^{Sch} - P_2^{(0)} = (P_{2,G} - P_{2,L}) - P_2^{(0)} = (0 - 1) - 0 = -1 \, p.u. \tag{15}$$

$$\Delta Q_2^{(0)} = Q_2^{Sch} - Q_2^{(0)} = (Q_{2,G} - Q_{2,L}) - Q_2^{(0)} = (0 - 0.5) - 0 = -0.5 \, p.u. \tag{16}$$

In (15) and (16), positive and negative signs are considered for the generation power and load demand, respectively. By considering bus 1 as the slack bus, the Jacobian matrix is as follows:

$$[J]^{(0)} = \begin{bmatrix} J_1 & J_2 \\ J_3 & J_4 \end{bmatrix}^{(0)} = \begin{bmatrix} \dfrac{\partial P_2}{\partial \delta_2} & \dfrac{\partial P_2}{\partial |\mathbf{V_2}|} \\ \dfrac{\partial Q_2}{\partial \delta_2} & \dfrac{\partial Q_2}{\partial |\mathbf{V_2}|} \end{bmatrix}^{(0)} \tag{17}$$

Solving (11), (12), and (17):

$$J_1^{(0)} = |\mathbf{V_2^{(0)}}||\mathbf{V_1^{(0)}}||y_{21}| \sin\left(\theta_{21} - \left(\delta_2^{(0)} - \delta_1^{(0)}\right)\right) \tag{18}$$

$$J_2^{(0)} = |\mathbf{V_1^{(0)}}||y_{21}| \cos\left(\theta_{21} - \left(\delta_2^{(0)} - \delta_1^{(0)}\right)\right) + 2|\mathbf{V_2^{(0)}}||y_{22}| \cos\left(\theta_{22}\right) \tag{19}$$

$$J_3^{(0)} = |\mathbf{V_2^{(0)}}||\mathbf{V_1^{(0)}}||y_{21}| \cos\left(\theta_{21} - \left(\delta_2^{(0)} - \delta_1^{(0)}\right)\right) \tag{20}$$

$$J_4^{(0)} = -|\mathbf{V_1^{(0)}}||y_{21}| \sin\left(\theta_{21} - \left(\delta_2^{(0)} - \delta_1^{(0)}\right)\right) - 2|\mathbf{V_2^{(0)}}||y_{22}| \sin\left(\theta_{22}\right) \tag{21}$$

By applying the primarily estimated quantities in (18)–(21), we have:

$$J_1^{(0)} = 1 \times 1 \times 10 \sin\left(90 - (0 - 0)\right) = 10 \tag{22}$$

$$J_2^{(0)} = 1 \times 10 \cos\left(90 - (0 - 0)\right) + 2 \times 1 \times 10 \times \cos\left(-90\right) = 0 \tag{23}$$

$$J_3^{(0)} = 1 \times 1 \times 10 \cos\left(90 - (0 - 0)\right) = 0 \tag{24}$$

$$J_4^{(0)} = -1 \times 10 \sin(90 - (0 - 0)) - 2 \times 1 \times 10 \times \sin(-90) = 10 \tag{25}$$

Therefore:

$$[J]^{(0)} = \begin{bmatrix} 10 & 0 \\ 0 & 10 \end{bmatrix} \tag{26}$$

$$\begin{bmatrix} \Delta P_2^{(0)} \\ \Delta Q_2^{(0)} \end{bmatrix} = [J]^{(0)} \begin{bmatrix} \Delta \delta_2^{(0)} \\ \Delta |V_2^{(0)}| \end{bmatrix} \Rightarrow \begin{bmatrix} \Delta \delta_2^{(0)} \\ \Delta |V_2^{(0)}| \end{bmatrix} = \begin{bmatrix} 10 & 0 \\ 0 & 10 \end{bmatrix}^{-1} \begin{bmatrix} \Delta P_2^{(0)} \\ \Delta Q_2^{(0)} \end{bmatrix} = \frac{1}{100} \begin{bmatrix} 10 & 0 \\ 0 & 10 \end{bmatrix} \begin{bmatrix} -1 \\ -0.5 \end{bmatrix} = \begin{bmatrix} -0.1 \\ -0.05 \end{bmatrix} \tag{27}$$

Finally, we can write:

$$\delta_2^{(1)} = \delta_2^{(0)} + \Delta \delta_2^{(0)} = 0 + (-0.1) = -0.1 \ rad$$

$$|V_2^{(1)}| = |V_2^{(0)}| + \Delta |V_2^{(0)}| = 1 + (-0.05) = 0.95 \ p.u.$$

$$\Rightarrow V_2^{(1)} = 0.95 < -0.12 \ rad$$

Choice (1) is the answer.

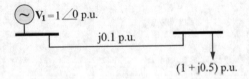

$V_1 = 1 \angle 0$ p.u.

j0.1 p.u.

$(1 + j0.5)$ p.u.

Fig. 10.8 The power system of solution of problem 10.15

Index

A

Accelerating factor (α), 85, 91
Active power, 4, 17, 18, 21, 91
Admittance, 2, 10, 15

B

Balanced three-phase power system
    impedance, 11, 32–34
      single-line diagram, 6, 22, 23
Base impedance, 20, 22
Base quantities, 5
Base voltage, 20, 24, 35, 79
Bundling, 37, 39, 41, 43, 46, 51

C

Capacitance, 39, 40, 43, 46, 47, 49, 50, 59
Capacitor, 76
Complex power, 6, 18, 21, 31
Conductance, 2, 15
Conductors, 37–41, 43, 44, 47–49
Conductors bundling, 37, 43
Consuming power, 4
Corona power loss, 37, 43
Current, 1, 2, 8, 11

D

DC load flow (DCLF), 85, 91
    active power, 86, 91, 92
    determine δ, 86, 92
    P_{12}, 86, 92, 93
    P_{G2}, 86, 87, 93, 94
    phase angle, 87, 94, 95

E

Economic load dispatch, 95, 96
    load demand, 88, 96
    power generation system, 96, 97
    total load demand, 88, 96, 97
Electric filed, 37, 43
Electric machine, 4, 19
Equivalent admittance of loads, 31
Equivalent impedance of load, 3, 17

F

Ferranti effect, 53, 59
Full-load transmission line, 53

G

Gauss-Seidel load flow, 85, 89, 91, 98
Generating reactive power, 4
Generation cost functions, 88, 97
Generator, 2–5, 15, 16
Geometrical Mean Distance (GMD), 46, 51
Geometrical Mean Radius (GMR)
    bundled conductors, 44
    conductors, 37–41, 43–50

I

Impedance, 2, 7, 11, 14, 15, 24, 32–35
Inductance, 38–41, 43–45, 47, 51
Inductor, 76
Instantaneous power, 3, 17

J

Jacobian matrix, 87, 94, 100

K

KCL, 16

L

Lagrange multiplier (λ), 88
Load bus (P-Q bus), 85, 89, 91, 99
Long transmission line model
    characteristic impedance, 56, 64, 65
    charging current, 56, 64
    open circuit, 56, 65
    short circuit, 56, 65
    transmission matrix, 64
Lossless transmission line, 54, 55, 57, 62, 66
Low-load transmission line, 53

M

Magnetic field, 37
Medium transmission line model
    charging current, 54, 60, 61
Motor, 4

N

Network admittance matrix, 97, 99
    diagonal components, 72, 80
    four-bus power system, 73, 82
    power system, 70, 73, 75, 77, 83

© The Author(s), under exclusive license to Springer Nature Switzerland AG 2022
M. Rahmani-Andebili, *Power System Analysis*, https://doi.org/10.1007/978-3-030-84767-8

Printed in the United States
by Baker & Taylor Publisher Services